Lecture Notes in Networks and Systems

Volume 582

The series "Lecture Notes in Networks and Systems" publishes the latest developments in Networks and Systems—quickly, informally and with high quality. Original research reported in proceedings and post-proceedings represents the core of LNNS.

Volumes published in LNNS embrace all aspects and subfields of, as well as new challenges in, Networks and Systems.

The series contains proceedings and edited volumes in systems and networks, spanning the areas of Cyber-Physical Systems, Autonomous Systems, Sensor Networks, Control Systems, Energy Systems, Automotive Systems, Biological Systems, Vehicular Networking and Connected Vehicles, Aerospace Systems, Automation, Manufacturing, Smart Grids, Nonlinear Systems, Power Systems, Robotics, Social Systems, Economic Systems and other. Of particular value to both the contributors and the readership are the short publication timeframe and the world-wide distribution and exposure which enable both a wide and rapid dissemination of research output.

The series covers the theory, applications, and perspectives on the state of the art and future developments relevant to systems and networks, decision making, control, complex processes and related areas, as embedded in the fields of interdisciplinary and applied sciences, engineering, computer science, physics, economics, social, and life sciences, as well as the paradigms and methodologies behind them.

Indexed by SCOPUS, INSPEC, WTI Frankfurt eG, zbMATH, SCImago.

All books published in the series are submitted for consideration in Web of Science.

For proposals from Asia please contact Aninda Bose (aninda.bose@springer.com).

More information about this series at https://link.springer.com/bookseries/15179

Irina Samoylenko
Editor

Advances in Management, Business and Technological Systems

Road Towards Sustainable Development

Springer

Editor
Irina Samoylenko
Informational Systems Department
Stavropol State Agrarian University
Stavropol, Stavropol Territory, Russia

ISSN 2367-3370 ISSN 2367-3389 (electronic)
Lecture Notes in Networks and Systems
ISBN 978-3-031-20802-7 ISBN 978-3-031-20803-4 (eBook)
https://doi.org/10.1007/978-3-031-20803-4

This Springer imprint is published by the registered company Springer Nature Switzerland AG
The registered company address is: Gewerbestrasse 11, 6330 Cham, Switzerland

Preface

Introduction

The concept of sustainable development has become particularly relevant due to the aggravation of economic, social and environmental problems. The main goal of sustainable development at the global level is humanity preservation, which requires the abandonment of nature-destroying technologies, structural transformation of the economy and changes in value orientations of public consciousness. This goal may be achieved by the balanced development of socio-ecological and economic systems at various levels. The conceptual foundations and instruments of sustainable development are becoming national and regional in nature.

The regional economy is characterized by the specific uneven socio-economic development of entities. This aspect is caused by the heterogeneity of production factors and interregional differences in solving social problems, such as unemployment, different income levels of the population and mechanisms of regional state support. At the same time, sustainable development of regions aims to implement measures for achieving economic growth, developing social sphere, improving the quality of life, enhancing the demographic situation and developing industrial, social and market infrastructure. In addition, regional sustainable development focuses on ensuring economic, food and environmental security in current conditions of digitalization.

The post-industrial transformation of national economy at an innovation stage of development requires research that is actual for emerging processes in a society in the format of sustainable development. Modern methodological tools for studying various aspects of sustainable development are extensive and differentiated. At the same time, a comprehensive assessment of the achieved level of regional development and its systemic socio-ecological and economic balance requires a combined use of various tools and methods for solving multidimensional tasks of sustainable development. Moreover, a number of methods are based on the use of statistical analysis to quantify the process of sustainable development, but a systematic approach to solving problems of sustainable development requires the use of expert surveys and other intuitive methods of obtaining initial information to build theoretical constructions and practical conclusions.

The improvement of tools for assessment and management of regional sustainable development should be directed at preserving development mechanisms, regardless of external and internal conditions and factors. Meanwhile, economic, social and environmental components become a subject to positive or negative effects from external factors, which necessitates comprehensive research in the field of management, business and technology towards sustainable development.

In this regard, this book includes studies of conditions and opportunities for achieving sustainable economic growth; problems of law and law enforcement, state and corporate governance in ensuring sustainable regional development; innovative entrepreneurship; the use of information and communication technologies for sustainable development; the role of social institutes for improving the quality of life.

Contents

Forecasting Methods as a Tool for Strategic Planning of Sustainable Development of an Agrarian Enterprise

Dzhannet A. Tambieva[1](✉) and Madina U. Erkenova[2]

[1] Stavropol State Agrarian University, Stavropol, Russia
tamjannet@mail.ru
[2] North Caucasus State Academy, Cherkessk, Russia

Abstract. Sustainable development of an agro-industrial complex (AIC) in the context of market economy requires active introduction of new concepts in strategic planning of business processes based on the latest forecasting tools. The forecasting of business processes in AIC is mainly represented by the individual models referring to informal expert reasoning. Global promotion processes for digital technologies in all economic areas require the creation of specialized systems of decision support based on the formalized forecasting models. This paper provides new approaches to forecasting socio-economic and climatic processes and/or systems in the context of strategic planning of AIC sustainable development. The analysis of dynamic specific characteristics of socio-economic and natural-climatic processes and/or systems has been implemented. The methods of probability theory, mathematical statistics and nonlinear dynamics (fractal analysis) were implemented for strategic planning of sustainable development of the AIC enterprises and the agriculture industry in general. Data mass analysis according to the key indicators of socio-economic and climatic processes and/or systems provided a conclusion about the presence of a "memory effect" in their dynamics. This fact makes it possible to form an adequate strategy of business processes and minimize the negative consequences of unfavorable events.

Keywords: Agro-industrial complex (AIC) · Forecasting · Strategic planning · Climatic and socio-economic processes and/or systems · Deterministic/stochastic process

1 Introduction

Strategic planning of AIC operation in general and its enterprises in particular, in the context of market relations, is impossible without studying the objective laws of cyclical dynamics of the corresponding socio-economic and climatic processes. Periodic economic crises of different depth and duration are the objective reality of the nature of market relations, including medium-term crises observed at the intervals of about once a decade, long-term—semisecular, so-called "Kondratievskie" and ultra-long-term—secular or, in other words, "civilizational".

I. Samoylenko (Ed.): AMBT 2022, LNNS 582, pp. 1–9, 2023.
https://doi.org/10.1007/978-3-031-20803-4_1

Along with the specified features of economic processes dynamics, the solution of the problem of strategic planning at the agricultural enterprises is impossible without forecasting the features of the corresponding natural and climatic processes, including the characteristics of the regional hydrological system (risks of drought and/or floods).

All this jointly is one of the most difficult tasks of the modern economy: the creation of an adequate mathematical tool for predicting the dynamics of socio-economic and climatic processes and/or systems, aimed besides at the purposes of strategic planning of AIC sustainable development and prevention and minimization of the negative consequences of the industry risks.

The efforts to solve this problem have actually led to the scientific paradigm replacement, appearance of new scientific trends, such as methods of nonlinear dynamics [1, 2], fuzzy logic [3], neural networks [4], etc. The classical forecasting models based on approximation methods, probability theory and mathematical statistics aimed at the whole class of problems turned out to be inconsistent with the required adequacy level. It should be noted that it is more typical for natural-climatic and socio-economic processes and systems.

Along with the well-known concepts of "deterministic/stochastic" system and/or process, the terms "deterministic chaos", "nonlinear system", etc. have been tightly incorporated in the modern science vocabulary.

The choice of forecasting method should be anticipated by the classification problem solution making it possible to determine the determinism level of considered system and/or process: to assess the tendency availability, possible cyclicality (quasi-cyclicality), periodicity (quasi-periodicity), etc.

In the context of market economy, AIC sustainable development needs the tools development for the business processes strategic planning based on forecasting of the key socio-economic and climatic processes and/or systems.

2 Materials and Methods

The research of any process and/or system involves data collection at an early stage. As a rule, the result of collecting such information comprises its representation in the form of a variational and/or time series [5]. The next most important stage is the process of data identification in the context of "determinism/stochasticity" is the analysis of relevance to one of the random variable distribution laws [6, 7]. The special aspects of the dynamics of socio-economic and natural-climatic processes and/or systems are the following: even if there are some individual exceptions, in general, these processes and/or systems do not comply with the normal or quasi-normal distribution law and poorly respond to the other random variable distribution laws.

One of these exceptions is shown below (see Fig. 1) – time series (TS) of consumer price indices (DP1) for the food products of one of the regions of the Russian Federation (TS DP1).

The statistical indicators have been calculated for TS DP1: minimum, maximum, range, mathematical expectation, dispersion, standard deviation, values of variation coefficients, asymmetry and kurtosis. A random variable distribution function has been

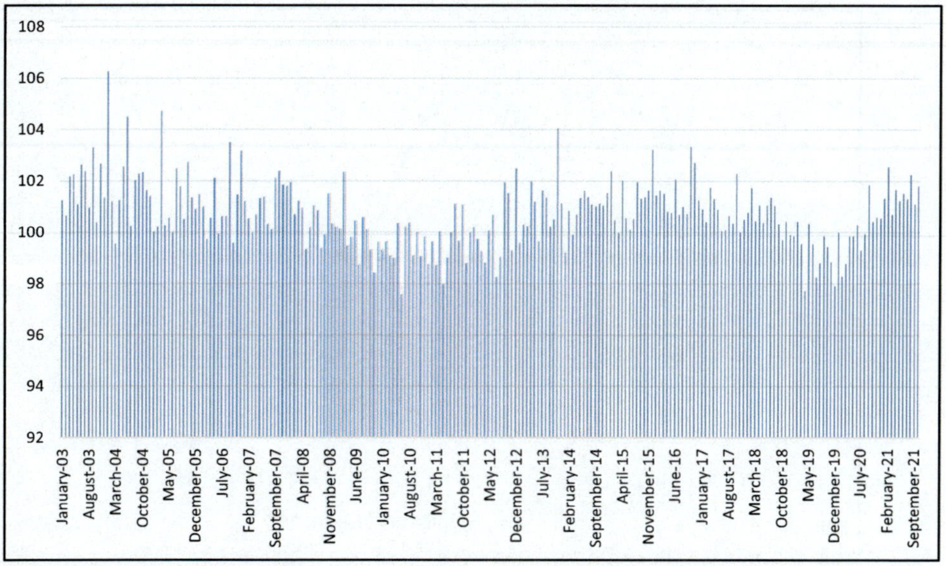

Fig. 1. Time series of consumer price indices for the food products of one of the regions of the Russian Federation—TS DP[1].

Table 1. Statistical indicators of time series (TS) DP1.

Statistical indicators	Value
Minimum (*min*)	97,59
Maximum (*max*)	106,26
Range (*R*)	8,67
Mathematical expectation (*MX*)	100,73
Dispersion (*DX*)	1,59
Standard deviation (*SX*)	1,264
Variation coefficient (*VX*)	1%

constructed (Fig. 2) using Sturges' formula [8]. The calculation results are shown in the Table 1.

Obviously, the classical statistical analysis has been reasonably informative for TS DP1. The tolerated forecast error does not exceed 1%. The forecast based on the mathematical expectation is adequate for this type of time series and can be used for the strategic planning purposes.

[1] Source: https://stavstat.gks.ru/folder/29846?print=1.

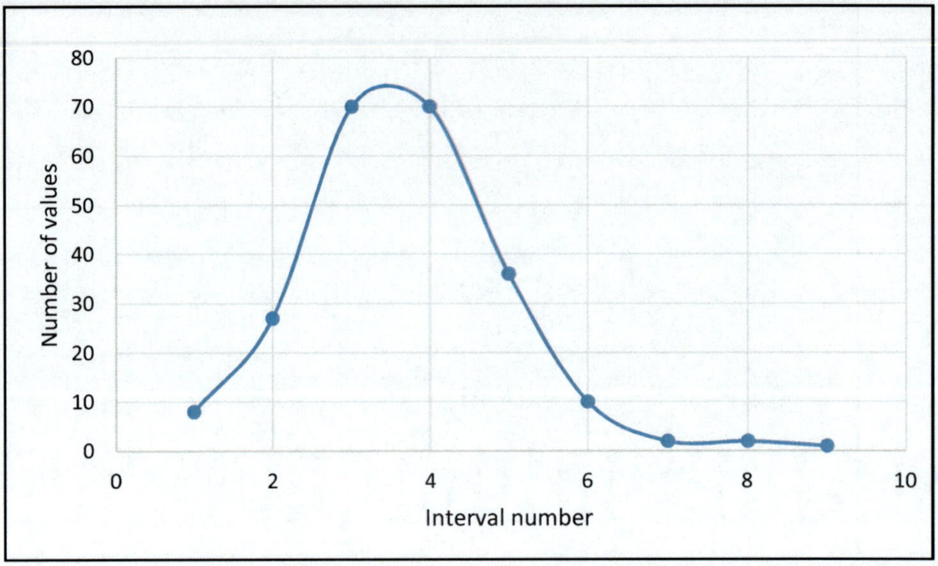

Fig. 2. Distribution function of random variable TS DP1.

However, the same approach for cereals productivity, presented in [9] in the territories of the modern Russian Federation from 1801 to 2008 has demonstrated non-compliance with the regular distribution law of a random variable (see Fig. 3 and 4). The similar result was obtained for the natural and climatic time series of average daily air temperature, water level in the rivers of one of the regions of North-Caucasus Federal District.

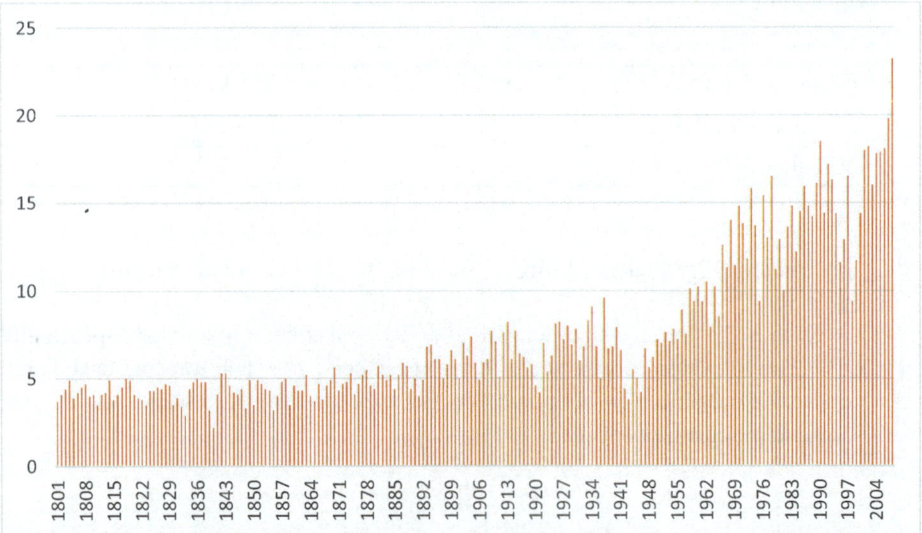

Fig. 3. TS of cereals productivity in the territories of the modern Russian Federation from 1801 to 2008.

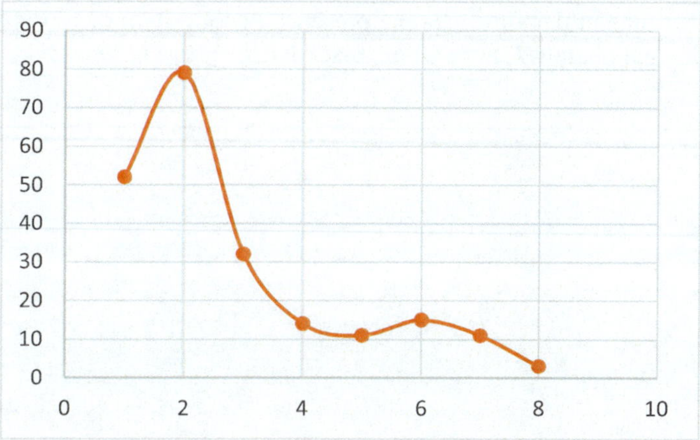

Fig. 4. Distribution function of random variable for the TS of cereals productivity in the territories of the modern Russian Federation from 1801 to 2008.

The investigation of TS for cereals productivity in the territories of the modern Russian Federation from 1801 to 2008 with the use of the method of R/S analysis [10–12] made it possible to qualify this series as a time series with "memory" [10–12]. The "memory effect" in the time series makes it possible to identify the cyclic component value (Fig. 5).

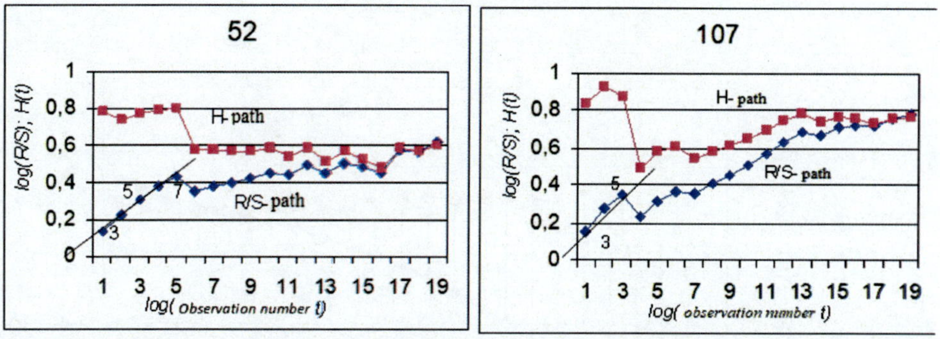

Fig. 5. The examples of quasi-cycles identified on the basis of R/S-analysis for the observations No. 52 and No. 107 for the TS of cereals productivity in the territories of the modern Russian Federation from 1801 to 2008. The memory depth is 7 and 5, respectively.

3 Results

The similar investigations have been implemented for the TS of daily observations of daily-average air temperature for the period from January 1, 2016 to December 31, 2016 in one of the regions of North-Caucasus Federal District (Fig. 6). The researches have shown that this TS also does not comply with the regular law of random variable distribution and demonstrates the presence of long-term memory (Fig. 7) – the values of Hurst exponent (H) are within "black noise" [10, 11].

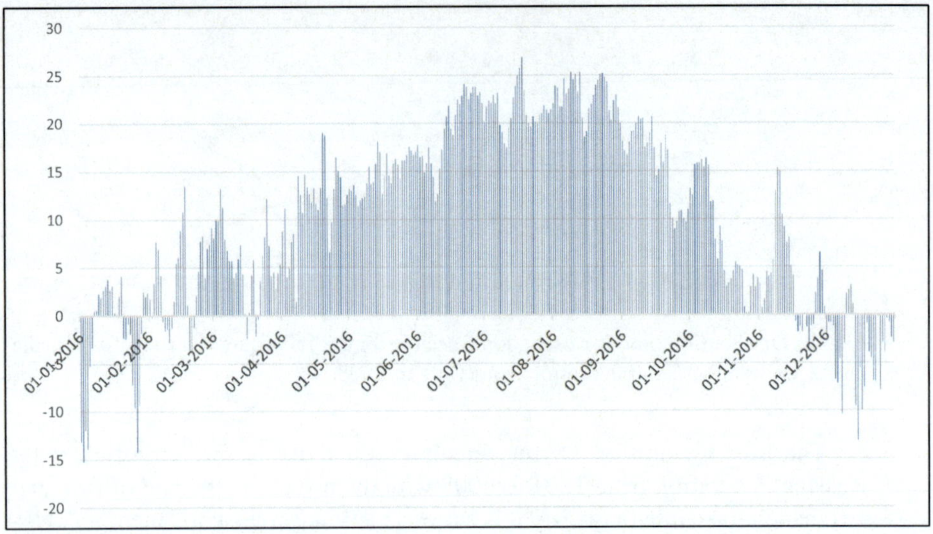

Fig. 6. Histogram of seasonal temperature variations (daily observations of daily-average air temperature for the period from January 1, 2016 to December 31, 2016 in one of the regions of North-Caucasus Federal District). *Source* (compiled by the authors and based on [12]).

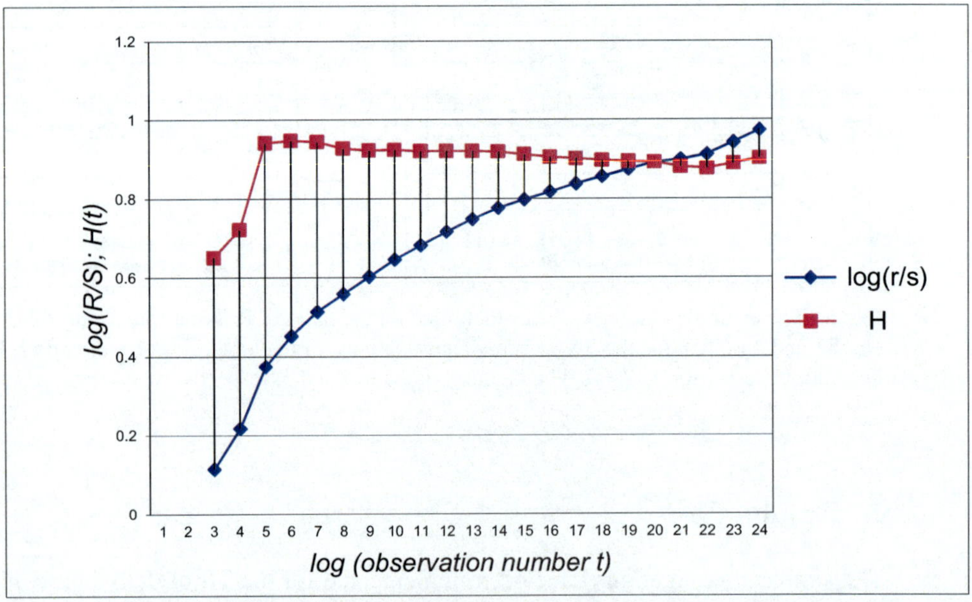

Fig. 7. Visual representation of R/S-analysis for a time series of seasonal temperature variations (see Fig. 3). *Source* (according to the authors' calculations).

The similar results have been obtained for the time series of the level of the Kuban River (Fig. 8 and 9) and Teberda River and other hydrological system indicators.

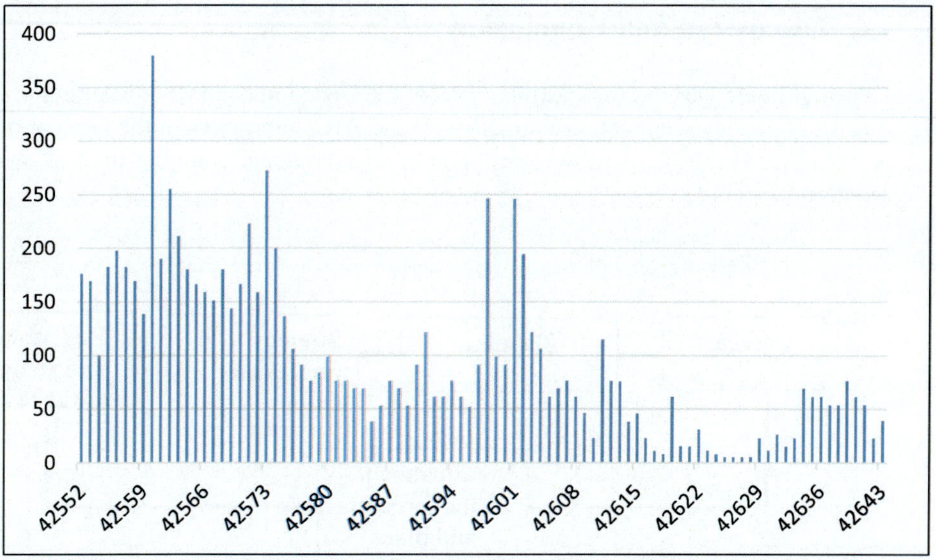

Fig. 8. Time series of daily observations of the water level in the Kuban River at the gauging station named after Kosta-Khetagurov from July 1, 2016 to September 30, 2016. *Source* (compiled by the authors and based on [13]).

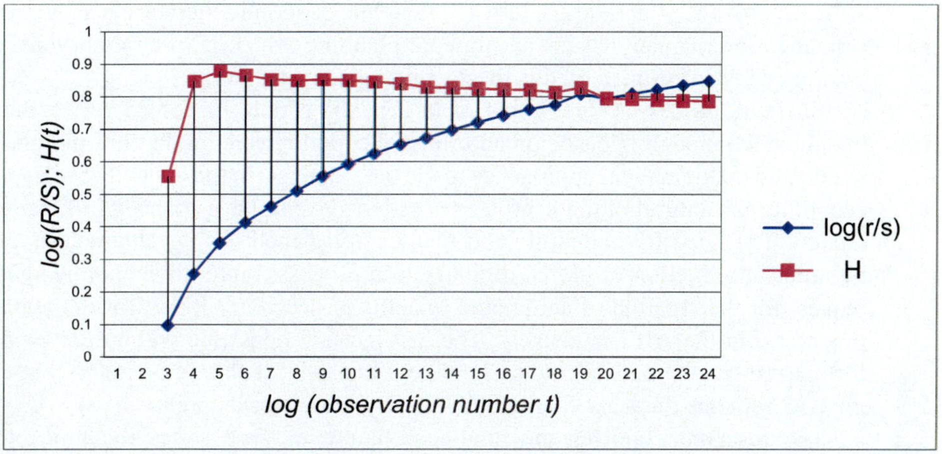

Fig. 9. Visual representation of R/S-analysis for a time series of the water level in the Kuban River (see Fig. 8). *Source* (according to the authors' calculations).

The classification problem solution at an early stage of data analysis makes it possible to assess the degree of determinism of considered socio-economic or natural-climatic process and/or system, to select an adequate predictive model for strategic planning purposes and, as a result, to develop an appropriate strategy increasing development sustainability of both specific elements and the industry in general.

4 Discussion and Conclusion

As highlighted above, forecasting is one of the key components of the strategic planning process for development sustainability of the AIC enterprises in the context of market economy. The Fig. 10 shows the "scheme of a strategic planning process at the AIC enterprises" [14].

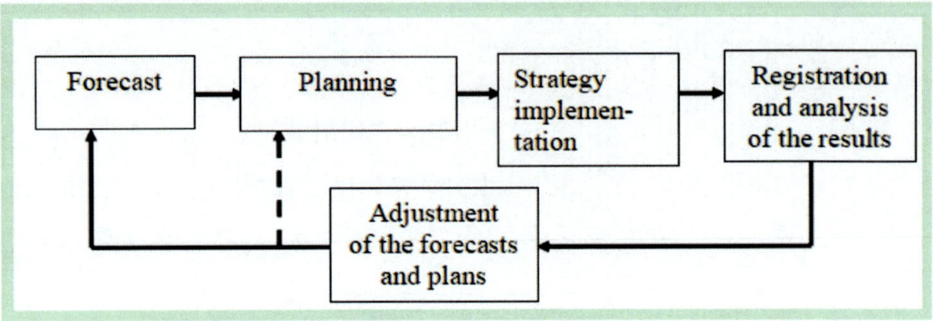

Fig. 10. The scheme of a strategic planning process at the AIC enterprises[2].

It is highlighted by Ternovykh K.S., Alekseenko A.S., Annenko A.S., Kuzmitskaia A.A., Ozerova L.V. and others [14, 15] that "the economic literature hardly provide the formalization of many issues of strategic planning which is in an embryonic stage at enterprises." Unfortunately, this thesis remains relevant so far.

Business processes forecasting of in the AIC should be considered in the context of subjective modeling based on an informal expert reasoning. In our opinion, the economic and mathematical approaches to the data analysis in the context of relevant socio-economic and natural-climatic processes and/or systems have not been sufficiently investigated. The need for economic and mathematical methods development is prompted by rapid introduction of global digitalization processes into all economic spheres, the request for the creation of automated systems of decision support for the management processes, in the AIC inclusively. The last fact is impossible without development of formalized (mathematical) methods of data analysis. It should be noted that currently the AIC remains the least "digitized" industry in all world countries.

The research confirms the nonlinear nature of time series dynamics of socio-economic and climatic processes and/or systems.

The distinction of the approach used in this paper is to give reasoning to application of the latest achievements in the field of modern prognostics (the methods of nonlinear dynamics, in particular,) to the purposes of business processes strategic planning in the

[2] Source: Kuzmitskaia, A.A., Ozerova, L.V.: Modern aspects of strategic planning organization in AIC. Manag. Econ. Syst.: An Electron. Sci. J. **63**(3), 13–25 (2014).

agro-industrial sector. It is aimed to support the development sustainability of both the industry in general and its components.

The implemented mass data analysis allows us to conclude about unambiguous determinism of natural and climatic processes and/or systems (see Fig. 7 and Fig. 9), characterized by the long "cycles" (quasi-cycles) and "long memory" [16]. In the context of fractal (R/S-analysis) methodology, the corresponding time series are describe as "persistent" [10]. At the same time, for the statistical data dynamics of crop productivity, the "memory depth" is limited above by a value corresponding to 11 [9].

Persistence characterizes an important quality of considered process and/or system and appears in maintaining the actual tendency for a long time. For example, an increase (decrease) in the past is more likely to lead to an increase (decrease) in the future.

Acknowledgements. The reported study was funded by RFBR, project No. 20-37-90102.

References

1. Feder, E.: Fractals. Mir, Moscow, Russia (1991)
2. Shuster, G.: Deterministic Chaos: Introduction. Mir, Moscow, Russia (1988)
3. Rutkovskaia, D., Pilinskii, M., Rutkovskii, L.: Neural Networks, Genetic Algorithms and Fuzzy Systems (Transl. from Polish by I.D. Rudinskii). Goriachaialiniia – Telekom, Moscow, Russia (2004)
4. Nazarov, A.V., Loskutov, A.I.: Neural Network Algorithms for Forecasting and Systems Optimization. NaukaiTekhnika, St. Peterburg, Russia (2003)
5. Kozlov, A.Iu.: Statistical data analysis in MS Excel: study guide. In: Kozlov, A.Iu., Mkhitarian, V.S., Shishov, V.F. (eds.) Infra-M, Moscow, Russia (2018)
6. Siegel, A.: Practical Business – Statistics. Viliams, Moscow, Russia (2002)
7. Kolemaev, V.A., Staroverov, O.V., Turundaevskii, V.B.: Probability Theory and Methods of Mathematical Statistics. Vysshaiashkola, Moscow, Russia (1991)
8. Sturges, H.: The choice of a class-interval. J. Amer. Statist. Assoc. **21**, 65–66 (1926)
9. Tambieva, D.A., Erkenova, M.U.: Methods of fractal and phase analysis for evaluating the cyclic component of one time series of crop productivity. Manag. Econ. Syst.: An Electron. Sci. J. **12**(106), 66–75 (2017)
10. Peters, E.: Chaos and order in the capital markets: a new view of cycles, prices, and market volatility. In: Gusev, V.I. Transl. from English; Romanova, A.N. (ed.), Mir, Moscow, Russia (Original work published 1991) (2000)
11. Peters, E.: Fractal Analysis of Financial Markets: Chaos Theory Application in the Investment and Economics. Internet-trading, Moscow, Russia (2004)
12. https://climate-energy.ru/weather/spravochnik/temp/climate_sprav-temp_3704702290.Php#anchor0. Accessed: 7 Apr 2021
13. https://allrivers.info/gauge/teberda-teberda. Accessed 07 Apr 2021
14. Ternovykh, K.S., Alekseenko, A.S., Annenko, A.S.: Planning at an enterprise of AIC. In: Ternovykh, K.S. (ed.), Kolos, Moscow, Russia (2007)
15. Kuzmitskaia, A.A., Ozerova, L.V.: Modern aspects of strategic planning organization in AIC. Manag. Econ. Syst.: an Electron. Sci. J. **3**(63), 13–25 (2014)
16. Beran, J.: Statistics for Long Memory Processes. Chapman and Hall, New York (1994)

Development of a Model for System Management of the Regional Entrepreneurial Ecosystem Factors

Oksana Nikolaevna Kolomyts[1], Inna Grigorevna Ivanova[2(✉)], Gaiane Agopovna Kochian[1], and Oksana Vladimirovna Vandrikova[3]

[1] Kuban State Technological University, Krasnodar, Russia
ksu_berimor@mail.ru
[2] Kuban State Agrarian University Named After I.T. Trubilin, Krasnodar, Russia
inna_ivanova_2010@mail.ru
[3] Kuban State University in Tikhoretsk, Tikhoretsk, Russia

Abstract. Acceleration of the global post-industrial transformations and expansion of spatial competition make necessary the business ecosystems formation at the meso-level as one of the advanced forms of economic policy implementation in the field of entrepreneurship. The authors developed an operator model in order to manage systematically entrepreneurial environment factors that determine entrepreneurial ecosystem development through "the nodal points of growth" in the relevant periods of development. This model application provides solving the problems of socio-economic development of Russian regions in an integrated way. Additionally, we show contributions of increasing regional stability, reducing social tension and ensuring employment growth.

Keywords: Ecosystem approach · Entrepreneurial environment · Entrepreneurial operations · Entrepreneurial ecosystem · Regional development · Entrepreneurial capital · Entrepreneurial activity

1 Introduction

Entrepreneurship plays an important role in ensuring the sustainable development of Russian economy for a number of reasons: from providing support to the social changes to innovation stimulation. Small business stimulates economic development by opening new enterprises, creating employment, being a kind of innovative ideas generator in high-tech economic sectors and contributing to the entrepreneurial potential fulfilment. For example, the share of small businesses in Japanese GDP (gross domestic product) is about 63%, in the USA—62%, in Malaysia—47%, and in Russia only 21%.

According to the International Monetary Fund, "currently, the world 90% of enterprises belonging to small and medium-sized businesses provide 63% of the global population with employment. The total share of small and medium-sized businesses in the European Union is 99.8% which ensure employment to 85% of the population" [1].

I. Samoylenko (Ed.): AMBT 2022, LNNS 582, pp. 10–17, 2023.
https://doi.org/10.1007/978-3-031-20803-4_2

However, despite the objective importance of entrepreneurship, the problem of successful functioning of contemporary Russian enterprises remains unresolved due to the artificial obstacles not allowing to fully using the small entrepreneurship potential.

According to the "Doing Business 2020" rating, from May 2018 to May 2019, 115 countries have implemented 294 reforms improving regulation in all areas, supporting entrepreneurial operations development and the norms limiting it. A number of quantitative indicators are used by "Doing Business" [2] to assess the level of favorableness of legal norms for entrepreneurial operations and property rights protection. These indicators are successfully used to data correlating for 190 world countries.

The indicator of favorable business conditions helps to assess the absolute level of a country efficiency in regulating business conditions for a certain period. It reflects the difference between each country's indicator and the best practices for each of the indicators among all countries.

The recent reforms have risen Russia from 31st to 28th place in the global ranking of ease of business operations. It is remarkable that Russia was ranked 120th eight years ago (Fig. 1).

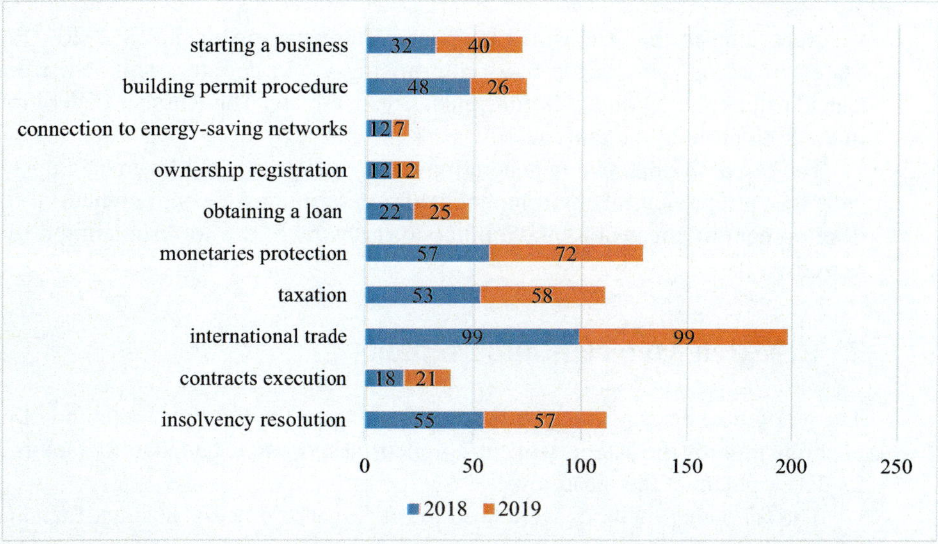

Fig. 1. Global rating of business operations in Russia during 2018–2019

The rating positively notes the reforms carried out in Russia in order to accelerate the process of electricity obtaining, simplify compliance with the tax requirements and strengthen the minority investors' rights protection by tightening corporate transparency [2].

The global entrepreneurship index, designed to measure qualitatively the entrepreneurial ecosystem in a national context, evaluates 101 economies in total, comprising r 95% of the global gross domestic product. The overall ranking takes into account a wide range of factors to create an overall index of "the best countries for entrepreneurship", which include innovation, competitiveness, infrastructure, labor skills, access to the capital and openness to business (Table 1).

Table 1. Global entrepreneurship index rankings 2020 [3]

Place in the rating	Country	2016	2017	2018	2019	2020
1	The USA	85,00	86,20	83,37	83,61	86,80
2	Switzerland	68,60	67,80	77,96	80,45	82,20
3	Canada	81,50	79,50	75,60	79,25	80,40
4	Denmark	71,40	76,00	74,06	74,31	79,30
5	The United Kingdom	72,70	67,70	71,29	77,75	77,50
…	……………	…	…	…	…	…
80	The Russian Federation	31,70	32,20	25,43	25,22	24,80

According to the "Ceoworld magazine Entrepreneurship Index 2020", the United States are recognized as the most enterprising world country, while Switzerland and Canada share the second and third places respectively. The Russian Federation is only in the 80th place in the ranking.

The research objective is generating of an operator model for managing systematically the entrepreneurial environment factors determining the entrepreneurial ecosystem development by means of "nodal points of growth" in the corresponding development periods.

2 Research Materials and Methods

The theoretical basis of the research is grounded on the dialectical method of cognition enabling to study the category of entrepreneurial ecosystem and identify factors affecting its development at the meso-level.

The following methods were used in the research process: abstract-logical, systematic, monographic, structural-logical and techniques of analysis and synthesis (grouping, typing, comparison). A number of applied methods and techniques are based on logical conclusions and calculated constructivism.

The information basis of the research was official statistical data, data from open electronic resources, official Internet sites of the International Monetary Fund, the rating "Doing Business-2020", "Ceoworld magazine Entrepreneurship Index 2020", etc., and the scientific publications on the investigated subject as well.

It is necessary to highlight the foreign researchers' works determining the concept of entrepreneurial ecosystems development. They are A. Tansley, G. Moore, D. Jackson, D.V. Eisenberg, R. Brown, K. Mason, B. Feld, E. Steam, Tian Xiu-Hua, Nie Qingkai, Zhang Xiaoren, Ding Lingand, Chen Xiangdong, Carayannis E., D.P. Campbell D.P., Zh. Karaslanian, K. Kakuda, S. Vatanabe, B. Mercon, D. Goktas, S. Durst and others.

However, currently in Russia the considered phenomenon is rather at the stage of scientific search and justification than in the area of practical use.

Despite the increasing number of studies on this issue various aspects, they do not provide a comprehensive view of the factors affecting the entrepreneurial ecosystems development at the meso-level, the ways of their management and evaluation of their impact on business environment. It confirms the relevance and timeliness of scientific research in the proposed field.

3 Results of the Research and their Discussion

Russian regional economy of is highly polarized—several relatively successful stand up against dozens of depressive or inertial regions, which is a large-scale problem of regional development. Russian researchers N.V. Zubarevich and S.G. Safronov [4] have paid their attention to the fact that in terms of development the differences between regions are often as significant as the differences between separate countries or groups of countries. However, there are key differences between municipal districts within each region.

We highlight that the policy implementation for small and medium-sized enterprises (SMEs) development keeps shifting to the regional level, because the territorial entities of the Russian Federation possess great potential significantly contributing to the socio-economic problems solving.

The lack of serious breakthroughs in the field of entrepreneurship development (with a few exceptions) demonstrates that the institutional conditions prevailing in a significant part of the regions have apparently exhausted their potential for business stimulation. Lack of development of Russian state institutions, incentive mechanisms, and investments necessary for business development do not meet modern needs and act as a kind of limiter for small business development in the regions of Russia.

Despite a large number of targeted programs and budget funds allocated to support entrepreneurship, it still does not significantly affect the level and pace of socio-economic territorial development.

In world practice, the concept of regional ecosystems [5–7] is actively used for entrepreneurial operations development policy. It uses an integrated approach to improving business conditions and creating a new look at regional area. The achievements in technologies and increasing globalization have changed the ideas about the best ways of doing business. It is believed that the idea of a business ecosystem helps companies to understand how to succeed in dynamic and constantly changing environment. The experience of industrially developed countries demonstrates that the ecosystem approach accomplishes the following points:

- stimulates formation of new forms of cooperation for solving economic, social and other problems;
- increases the level of innovation of small and medium-sized business enterprises;
- provides the opportunities to enter new markets;
- supports effective cooperation and knowledge, professional experience and skills exchange;

– allows using the territorial potential to the full extent, which contributes to modernization acceleration of the regional economy.

Within the framework of the ecosystem approach to entrepreneurship investigation, it is of great importance to take into account the impact of a specific space-time continuum in all its diversity on the entrepreneurial processes development. This aspect importance is especially significant for our country due to its huge territorial extent and complex administrative and territorial structure. It this context, the regions of the Russian Federation differ from each other not only in economic, industrial, natural resource parameters, but also often have different socio-cultural, institutional environment, growth reserves, and specific problems.

Each region is an independent entity of development, implementing its own strategy and providing support to small enterprises. The need to concentrate such support at the regional level is subject to independence expansion and background for the authorities' more effective actions as well.

The regional entrepreneurial ecosystems may become one of the promising forms of economic policy implementation in the field of entrepreneurship. It will allow solving the problems of socio-economic development of the territories, contributing to improving their sustainability, reducing social tension, ensuring employment growth, etc.

The scientific thought evolution determines a controversial nature of the concept "entrepreneurial ecosystem". Generalization, analysis and systematization of various points of view of economic science representatives [8–10] have allowed us to propose the following interpretation: entrepreneurial ecosystem can be defined as a complex of interrelated subsystems and elements of entrepreneurship as well as the environment of their functioning and interaction, contributing to entrepreneurial activity activation and entrepreneurial capital strengthening.

Accordingly, the entrepreneurial ecosystems possess the following basic features (Fig. 2).

The favorable business environment supports entrepreneurial ecosystems effective functioning and development ensuring effective implementation of entrepreneurial operations aimed at increasing of business ecosystems competitiveness.

We will assume the entrepreneurial environment (EE) as a totality of all conditions for entrepreneurial operations implementation, contributing to the innovations creation and implementation, providing the entrepreneurs with new opportunities, attracting investors and leading to the entire ecosystem growth.

Considering the entrepreneurial processes dynamism, we admit that emergent effect presentation in the form of continuous reproduction of a favorable business environment is more substantial. The additional emergent effects should also be considered from the perspective of the studied entrepreneurial ecosystems functioning and development, but already oriented to the producers, consumers, society and society as a whole [11].

One of the conditions for successful entrepreneurship development is formation and development of a relevant infrastructure as a system with a certain emergent effect in accordance with a complex of systemic principles [12]. The state should create and regulate entrepreneurial ecosystems development by providing favorable conditions for their operations.

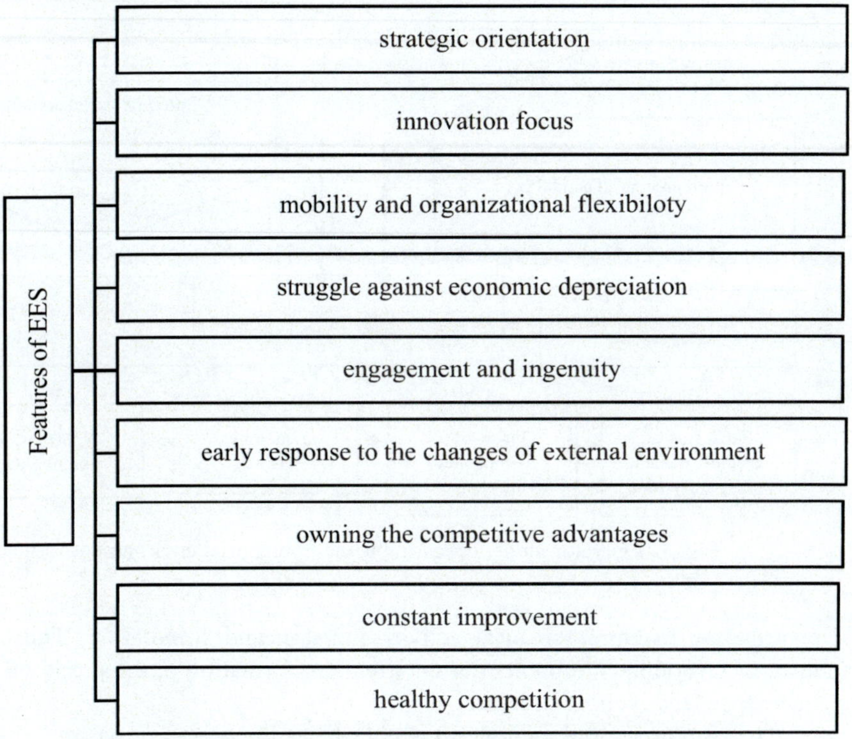

Fig. 2. Features of entrepreneurial ecosystems (EES)

This approach allows us to consider the operations of entrepreneurial ecosystems, supporting the organizations, state and other structures not in isolation, but in a system whose effectiveness is influenced by a complex of factors determining regional entrepreneurship current and projected development.

It should be highlighted that both the factors and the degree of their influence can change in time and geographically. Thus, a situation systematic monitoring and evaluation is a condition of great importance [13, 14].

Evaluation of these factors impact is possible with the help of a systematic analysis of regional business environment, which allows characterizing the existing conditions of entrepreneurial operations.

Entrepreneurship effective performance is impossible without taking into account the changes of territorial socio-economic environment, significantly influencing entrepreneurial operations strength.

Thus, a territory possesses dominant influence on small business subjects and objects and their economic interaction [15]. This fact causes the need to build an operator model ($F: X \rightarrow Y$) of the mechanism of favorable entrepreneurial environment formation. Its theoretical visualization is presented in the Fig. 3.

The economic and social functions are performed within the framework of the regional system, and the increase of entrepreneurial capital creates a multiplier effect,

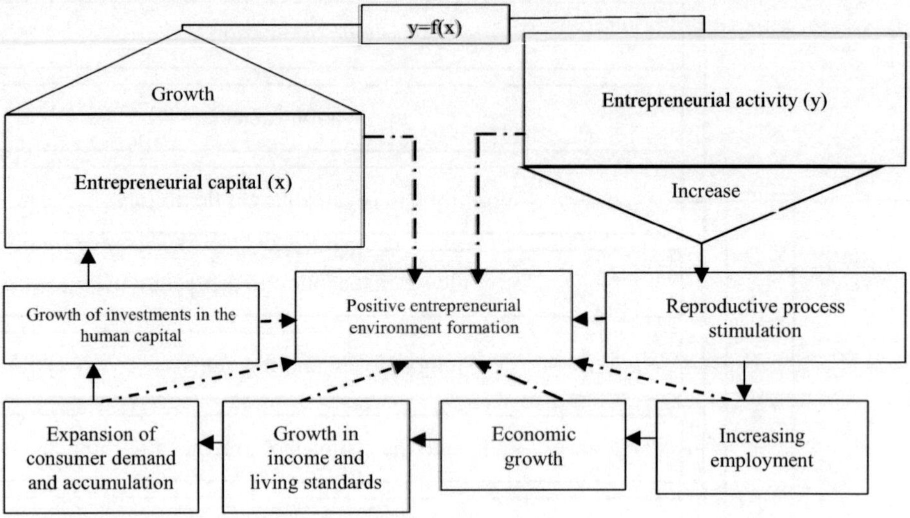

Fig. 3. Operator model of favorable entrepreneurial environment formation

contributing to entrepreneurial activity increase and stimulation. Thus, the mechanism of favorable entrepreneurial environment formation in the proposed model is an interdependent cyclical structure.

The entrepreneurial capital (x) is defined as the region provision with the factors contributing to new enterprises creation and the level of entrepreneurial activity increase (y), which has a positive impact on the region socio-economic condition. The regions with a higher level of entrepreneurial activity demonstrate a higher level of production and productivity.

A more complete model of this type allows obtaining diverse characteristics of business conditions and describe the dynamics of changes occurring in regional socio-economic environment.

4 Conclusion

The proposed model provides systematically the entrepreneurial environment factors determining the entrepreneurial ecosystem development by means of "nodal points of growth" in the corresponding periods of development.

The developed model will help the public authorities in assessing state support effectiveness, analyzing the positive and negative results of state regulation of small and medium-sized businesses development, and in creating effective forward-looking state policy.

Acknowledgement. The research was implemented with the financial support of Russian Foundation for Basic Research (RFBR) and Krasnodar territory in the framework of the scientific project 19-410-230041.

References

1. Report on the issues of global financial stability. Analytical summary. https://www.imf.org/ru/Publications/GFSR/Issues/2019/10/01/global-financial-stability-report-october-2019. Accessed 23 Mar 2020
2. Doing Business-2020: Country rating. https://russian.doingbusiness.org/ru/rankings. Accessed 25 Mar 2020
3. Global entrepreneurship index rankings: https://thegedi.org/global-entrepreneurship-and-development-index/. Accessed 26 Mar 2020
4. Zubarevich, N.V., Safronov, S.G.: Inequality of socio-economic development of Russian regions and cities in the 2000s: growth or decline? Soc. Sci. Modernity **53**(6), 15–26 (2013)
5. Armington, C., Odle, M.: Small business: how Many Jobs? The Brookings Rev. **1**(2), 14–17 (1982)
6. Feld, B.: Startup Communities: Building an Entrepreneurial Ecosystem in Your City, 224 p. Wiley (2012)
7. Isenberg, D.: How to start an entrepreneurial revolution. Harvard Bus. Rev. 41–50 (2010)
8. Jackson, D.J.: What is an Innovation Ecosystem? National Science Foundation, Arlington, V.A. (2011)
9. Mason, C., Brown, R.: Entrepreneurial Ecosystems and Growth Oriented Entrepreneurship, 38 p. The Hague, Netherlands (2014)
10. Stam, E.: Entrepreneurial ecosystems and regional policy: a sympathetic critique. Eur. Plan. Stud. **23**(9), 1759–1769 (2015)
11. Baidakov, A.N., Grigorian, A.R.: System aspects of innovative business infrastructure formation. Multi. online Sci. J. Kuban State Agrarian Univ. (124), 1293–1304 (2016)
12. Gladilin, A.V., Popov, M.N., Kolomyts, O.N.: Analysis and Evaluation of the Socio-Economic Environment for the Regional Investment Projects Implementation: Monograph, 177 p. Moscow (2013)
13. Velinov, E., Ponomarev, M.: Organizational development of Russian SMEs: current trends. J. Eastern Eur. Central Asian Res. (JEECAR) **3**(2), 10 (2016)
14. Kolomyts, O.N., Ivanova, I.G., Popov, M.N.: Regional economic development based on the effective functioning of entrepreneurial environment. Azimuth Sci. Res.: Econ. Manage. **8**(26), 176–179 (2019)
15. Kolomyts, O.N., Prokhorova, V.V., Ivanova, I.G.: Advantages and prospects of creating regional high-tech agrarian clusters. Indo Am. J. Pharma. Sci. **6**(3), 6881–6885 (2019)

Achieving Sustainable Growth of the Grain Product Subcomplex Production on the Basis of the Formation and Development of the Integrated Structures

I. A. Rodionova, V. V. Toropova, O. A. Vasilyeva, M. A. Bolokhonov(✉), and E. G. Afanasyeva

Saratov State Agrarian University named after N.I. Vavilov, Saratov, Russia
bolohonov@list.ru

Abstract. The article considers the main directions of improving the efficiency of the Russian and regional grain product market on the basis of the formation of integrated structures. In order to present the main trends and general directions of the grain market development, the technique of expert polling of specialists in the sphere of agriculture of different professional groups on the basis of ratings and personal opinions of respondents was applied, allowing to determine the main directions of the grain product market development. The structure of respondents' answers and the distribution of their preferences among the spheres of activity were analyzed. Based on the assumption that the efficiency of the grain product market is influenced by the organization of production and production relations on a par with the methods of production, the parameters of enterprises, that are important when uniting into a cluster, were singled out according to their priorities. The priority directions of the rational use of the land resources are considered, the recommendations of solving the main problems and imbalances of the plant growing industry, in particular the containment of the share of sunflower crops are grouped. The directions of increasing the efficiency of grain product market, functioning on the basis of the mechanism of integration and concentration of production, have been proposed, the main directions of agrarian politics, which allow to achieve the main objective of the regional grain product market functioning, namely, providing the population with the products of own production, are grouped.

Keywords: Grain products · Market · Cluster · Poll · Rating

1 Introduction

The zoning of agro-industrial production and the internal organization of the agrifood market for certain types of products leads to the decentralization of the grain product market. Decentralization of the agrifood market causes the necessity of the convergence of agricultural production, processing enterprises and trade organizations, which

I. Samoylenko (Ed.): AMBT 2022, LNNS 582, pp. 18–27, 2023.
https://doi.org/10.1007/978-3-031-20803-4_3

contributes to the development of integration contacts in the agrifood market. The integration is necessary for the successful development of the grain product market [1]. It optimizes the supply of the agricultural products, providing the processing industry with raw materials, and maintains the necessary level of product consumption. Moreover, the integration processes are currently developing at the global markets as well [7], where Russia has been involved increasingly in recent years as an exporter of grain products [14].

In the countries with developed market economies, one of the most promising ways of economic development and increasing the competitiveness of the state is a cluster policy [10, 11]. Economists and economic policy planners are increasingly starting to focus on the topic of local industrial agglomeration and specialization as the most relevant at the moment. Enterprises in the sector through innovations support business adjustability and foster endogenous regional development in an expanding global market [2].

The objective of the research is to substantiate the necessity of the development of the integrated structures in the grain product subcomplex of the agroindustrial complex of Russia, as well as to determine the most relevant and popular directions of improving the efficiency of the grain product subcomplex and the grain market, which will ensure sustainable growth of production by increasing the economic efficiency of production and sales of grain. The stated goal was reached by determining the main directions of development of the regional grain product market on the basis of expert opinion of agricultural specialists and the research of principles of optimization of development of grain product subcomplex on the basis of formation and development of integrated structures. The importance of the development of integrated structures is supported among scientists and practitioners in the sphere of agricultural production, which was confirmed by the conducted poll.

2 Materials and Methods

The study was based on informational analytical materials, the legislation of the Russian Federation, official statistical data of the Federal State Statistics Service and the Ministry of Agriculture of the Russian Federation and Internet resources.

The article employs both universal and general scientific approaches to the analysis of the problem under study: dialectics, abstraction, deduction, induction, analysis and synthesis. Specific methods of the conducted research: analysis and synthesis, economic-statistical method, method of a comparative analysis and expert assessments.

The economic ties, underlying the formation of clusters can be based on a number of long-term contracts and operate through vertical-horizontal interactions between enterprises. Grain product clusters, as geographically concentrated companies and organizations, may include both processing enterprises and agricultural organizations, or even companies producing mechanization equipment and chemicals for agricultural production. Agrarian integration of regions specifies the industry specialization which is connected with the most optimal climatic conditions for the production of one or the other type of grain products.

Saratov region and Russia on the whole have a great potential for the development of the grain product market, but in order to raise its efficiency it is essential to determine the directions of its development and the list of required measures should be taken by the regulating bodies and the participants of the market.

To identify the areas of the efficiency of the grain market improvement, we conducted a poll for the specialists of the agro-industrial complex, which makes it possible to understand the main trends and overall directions of development.

We formulated a 12-item questionnaire, including two questions about the respondent's characteristics.

Most of the questions contained fixed answer options, for example question 1 had options, 5 questions besides the ones listed suggested their answer options, in three questions one could choose several options, practically all questions with the exception of four suggested the option " I find it difficult to answer".

The poll was conducted among 4 main groups of respondents employed in the agricultural sector (50 people totally): academic staff of the institute, which deals with the problems of agro-industrial complex, employees (specialists) of agricultural enterprises from the rural areas, teachers and students of agrarian higher educational establishments.

3 Results

3.1 Grouping the Results of the Poll

The results of the poll were arranged in the form of figures and diagrams with the grouping of the respondents' answers.

The results of the survey are presented below. Answers to the first question "Do you think that the efficiency of the grain product market is insufficient and needs to be improved?" were as follows: 24% "no", 67% "yes", 9% found it difficult to answer.

Answers to the second question were distributed as follows: 27 people consider that the annual growth of agricultural production is possible by 3–5%, 7 people - by 6–7%, 5 people. - by 8–10%, 6 people - by 11–15%. Only 3 respondents stated that there will be a decline in production, 2 respondents found it difficult to answer.

The majority of respondents (59%) assume that over the next 3–5 years, grain product market volume will not change, 26% chose the option "decline" and 16% of respondents decided that it will increase.

A structure of responses to the question, "What sector of the grain market is a low-performing and needs improvement?" is presented (Fig. 1). The majority of respondents consider that it is necessary to improve the efficiency of agricultural production, while a fifth of the respondents are in favor of increasing the trade service efficiency.

Significant interest lies in the prioritization of enterprise parameters' types, which are relevant when merging into a cluster, to which it was proposed to assign the appropriate meaning (Table 1).

Respondents placed profit increase on the first place, cost minimization on the second, human resources (managerial and working) and resource intensity on the final places. In paragraph 3.3 there is an interpretation of the values of these parameters used in the evaluation of enterprises, pretending to be united into a cluster.

The majority of respondents consider, that at the current moment it is necessary to remove administrative barriers (53%), about a third of respondents (29%) suggested, that

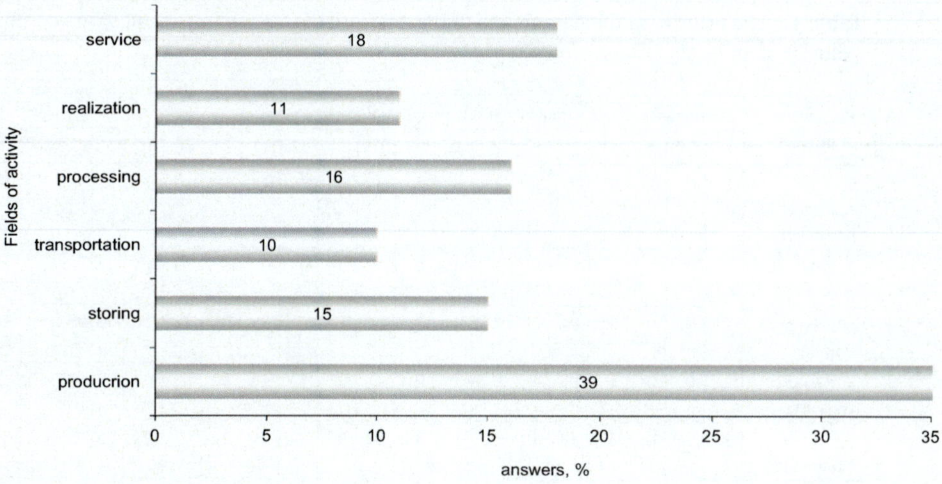

Fig. 1. Structure of answers to the question: "What area of the grain market is low efficient and requires improvement?"

it is necessary to increase the state support, 17% chose the reduction of energy tariffs, a tenth part of respondents are confident, that it is necessary to increase the availability of loans and leasing, and 5% demanded an increase in the state control.

Then follows the distribution of respondents' preferences among the areas of activity, that need to be given the greatest attention in order to improve the efficiency of the grain market (Fig. 2), where the first place is occupied by the technical re-equipment, and then comes the marketing activities, seed production and research activities, which indicates the necessity of focusing on the development of production methods and techniques, which ensure the efficiency of the grain product market.

The opinion on whether it is necessary to support the production and processing of grain in small forms of management (Peasant farm, Personal subsidiary farm) was ambiguous - 45% said that this was necessary, 37% refused to support them, 18% could not make a decision on this question.

According to the third part of the respondents, it is possible to increase the production and processing of grain products through the introduction of the new varieties (31%), implementation of fertilizers (23%), increasing the integration (21%), concentration (19%), specialization (15%). The results demonstrate that, according to the opinion of experts, the efficiency of the grain product market is influenced by the organization of production and manufacturing links on a par with the production methods.

Below there is a structure of answers to the question: "How to improve the marketing of grain and grain products? (Fig. 3). Almost half of the respondents suggested creating production clusters, more than a third supported the development of cooperation, a little less than a fifth considered that it is necessary to develop collective farms.

In summing up the results it is necessary to take into account the 5% statistical error in the calculations and, consequently, the real indicators, characterizing consumer preferences may fluctuate within about 5%.

Table 1. The priority of the parameters of the enterprises that are important when combining into a cluster, people.

Parameters of products	Significance of the parameter, place, people																	
	1	2	3	4	5	6	7	8	9	10	11	12	13	14	15	16	on average	summarized
1. Managerial personnel	1	0	1	0	0	1	3	0	0	0	2	2	12	18	10	0	12,70	14
2. Production personnel	0	0	1	2	1	0	1	0	5	1	0	0	0	8	19	12	13,34	15
3. Personnel training	2	0	0	3	0	0	2	0	1	1	1	0	3	0	16	21	13,46	16
4. Resource capacity	2	0	0	1	0	1	0	3	0	4	0	13	14	12	0	0	11,64	13
5. Profitability	3	3	0	6	18	10	1	1	3	0	0	1	2	2	0	0	5,82	5
6. Stability	3	2	1	3	7	1	4	4	12	9	1	1		1	0	1	7,48	9
7. Production volume	2	1	1	0	1	8	9	18	6	0	0	1	2	1	0	0	7,46	8
8. Production structure	1	2	4	5	1	6	15	6	1	0	0	1	1	4	0	3	7,38	7
9. Consumption structure	0	1	6	4	7	16	9	1	6	0	0	0	0	0	0	0	5,84	6
10. Equipment	1	0	0	0	0	2	0	0	12	18	13	0	0	2	0	2	10,08	10
11. Location	0	0	0	2	0	1	0	6	0	0	12	15	8	1	2	3	11,40	12
12. Expanding possibilities	2	0	0	0	0	1	0	3	0	14	17	11	2	0	0	0	10,34	11
13. Market growth	3	5	20	5	1	1	2	2	2	0	0	3	0	1	1	4	5,62	3
14. Quality improvement	2	2	4	19	4	2	4	3	1	3	3	1	2	0	0	0	5,76	4
15. Cost minimization	3	23	8	0	9	0	0	0	1	0	1	1	1	0	2	1	4,18	2
16. Increase in profit	25	11	4	0	1	0	0	3	0	0	0	0	3	0	0	3	3,50	1

3.2 Directions of Increasing the Grain Product Market and Grain Product Subcomplex Efficiency

Practically all the experts pointed out that the efficiency of the grain product market functioning is not high enough and demands measures aimed at increasing it. Besides, it follows from the experts' opinions, that one should not expect the situation to improve in a short period of time. The utmost attention should be paid to the reduction of administrative barriers and the increase of the state support in the sphere of grain production. As a result, we can identify a number of steps, that need to be taken at the level of enterprises of the grain product subcomplex in order to improve the efficiency of the grain market:

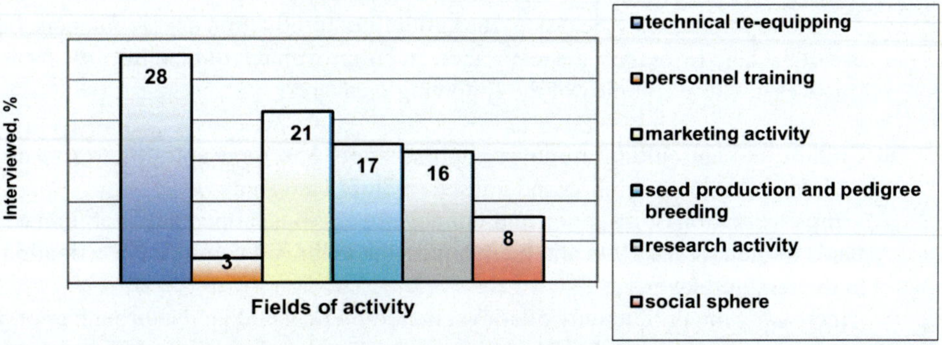

Fig. 2. Distribution of respondents' preferences among the areas of activity that need to be given the greatest attention for the improvement of the grain market efficiency

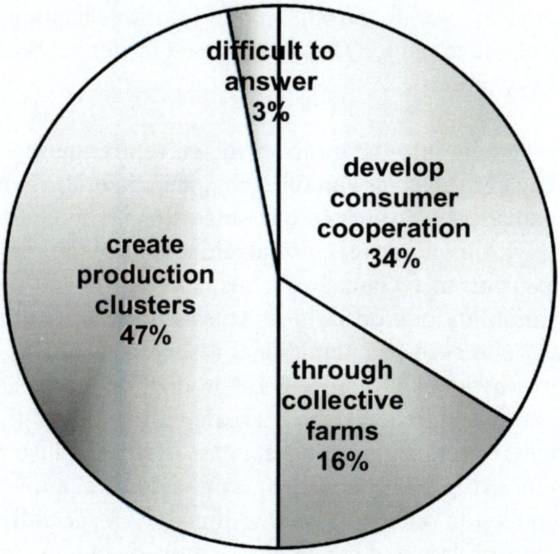

Fig. 3. Structure of answers to the question: "How to improve the marketing of grain and grain products?"

– Introduction of new technologies;
– Increasing of the production organization;
– Strengthening of the integration processes;
– Production consolidation;
– Creation of the agricultural clusters.

The implementation of these measures will enable to increase the efficiency of the grain market, increase the volume of production, decrease the transaction costs, increase the competitiveness of the grain product subcomplex enterprises and their products, and increase the level of satisfaction of the needs of consumers of grain and grain products.

An important strategic task of the agricultural production development is the rational use of the land resources, the restoration and improvement of land fertility. Its realization involves a whole complex of the following measures:

- return into agricultural turnover approximately 850 thousand hectares of arable land withdrawn from circulation and unused (mainly land shares);
- implementation of irrigation and drainage operations, aimed at restoration of irrigated lands to the level of 1990 and maintaining the technical condition of irrigation systems at the required level;
- Increasing the productivity of arable, including irrigated lands, through a set of special measures (agrochemical, meliorative and cultural-technical, including environmental forest protection, anti-erosion) [13] and improving the structure of sown areas in accordance with the requirements of a rational system of crop rotations, namely, increasing the area of perennial grasses, especially in irrigated lands, application of productive high-quality seed material, introduction of intensive technologies;
- Development and transition to science-based farming systems adapted to the local soil and climatic conditions.

The involvement into the turnover the currently unused land areas, the reconstruction of the irrigation systems and further expansion of the irrigated land areas, the whole range of reclamation measures - are an important resource for the growth of agricultural production. Hydromeliorative, cultural and technical measures are aimed at the restoration and expansion of reclamation lands. Increasing soil fertility will smooth out the significant variability of crop yields in Russia. Instability of agricultural productivity in Russia is also observed in international researches [12]. However, this problem is no longer a purely national one, since grain production in Russia has an impact on the food security of many countries [3]. Such a scale will not allow to smooth down the negative factors of yield reduction only at the expense of the risk insurance [8], a more systematic approach to increasing the stability of crop yields is required.

Grain production remains the main direction of specialization in Russia's crop production. To consolidation of the country's position on the grain market it is necessary to continue increasing grain production, especially to increase the stability of gross yields, so that in the future they would not fall under 130 million tons, taking into account increasing exports. Increase in the share of produced crops should be due to the expansion of sown areas, their full and efficient use, increasing yields, improving the placement of the species composition, increasing focus on the cultivation of the more productive crops, the use of high-quality seeds and plant protection measures against pests and diseases. In addition, an extra factor contributing to an increase in the export of grain can be a reduction of domestic consumption by reducing the use of grain for livestock feed in "pure form".

Taking into the consideration the growth of grain production, the state of grain elevator and storage facilities and their location, along with the possible improvement of the financial situation of the farms, it can be expected that a number of regions may experience a certain shortage of capacities. It is necessary to provide the reconstruction, and possibly a new construction to replace outdated capacities in those regions where the depreciation rate exceeds 70% on the currently existing production areas.

It is also important to increase the areas under leguminous crops, perennial leguminous grasses to solve the protein problem in forage production, as with the purpose of restoring soil fertility, because leguminous crops are the best predecessors. As an example, in the Saratov region over 17% of the cultivated area is occupied by the sunflower. The size of cultivation of this crop exceeds the possible limits, since in the rotation the sunflower should be re-sown not earlier than once every 7–8 years. Further expansion of the areas under it is not reasonable. The reduction to a rational level is required. At the same time, it is preferable to expand cultivation of the other oil-bearing crops, such as mustard, soybeans and other oil-bearing crops, the crops of which have been reduced and even liquidated in some areas due to the lower level of marketability and competitiveness compared to the sunflower. One of the directions of crop production development should be the expansion of the variety of cultivated species that are in demanded on the grain market and possess a middle-income potential. In general, the main ways of improving the efficiency of the grain product market functioning are presented in the figure below (Fig. 4).

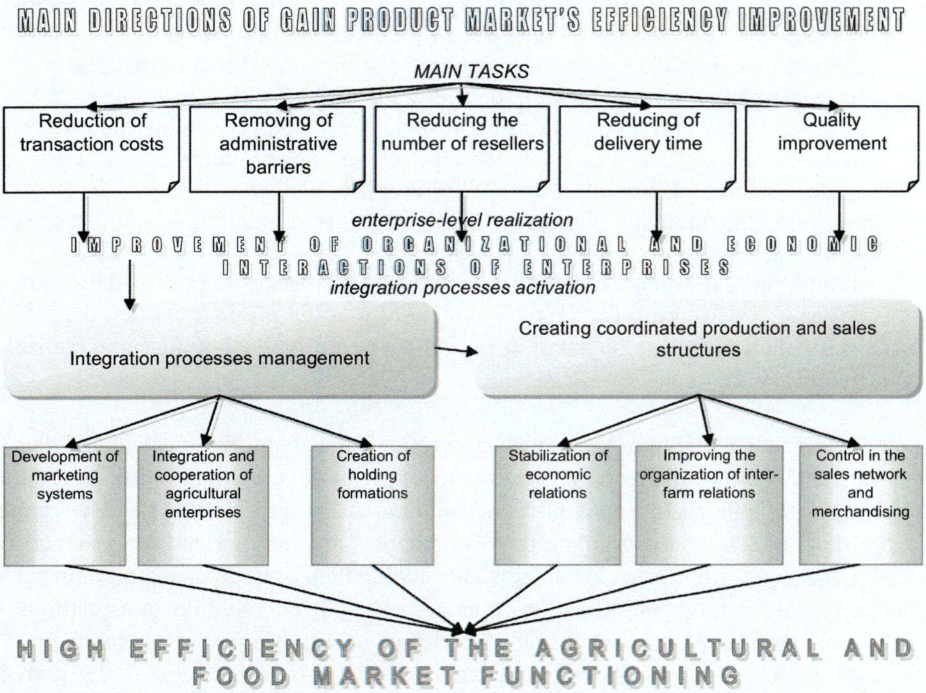

Fig. 4. The main directions of improving the efficiency of the grain product market

The strategic objective of the agrarian policy, realized in Russia, should be the creation of the essential conditions, providing the sustained functioning of the agro-industrial complex, aimed at the financial rehabilitation of the industry and the solution of its problems. Russia is currently one of the world' s largest grain producers and exporters. Foreign trade of agro-industrial complex products has a significant impact on the manufacturers' income, and the internal policy of import substitution in the conditions of sanctions directly and implicitly stimulates national production [5]. A significant part of grain products is already produced on private farms. Such structural and institutional changes imply shifting the focus of agrarian policy towards the stimulating of the minor forms of farming, as it has gradually happened in the USA. [4]. Nevertheless, in our opinion, such support should be based not on the criteria of expanding the production by any price and supporting the inefficient farms [6], but on the stimulation of expanded reproduction at the leading enterprises.

4 Discussion and Conclusion

The main task of the agricultural policy is to provide the population with products of own production. Foreign science and practice confirms the necessity of state regulation of agrifood markets [9]. On the basis of the results of the conducted poll, we outlined the most urgent directions of the agrarian policy:

ensuring of the state support for the agriculture at both the national and regional levels;
formation of regional food funds, seed funds, forage grain fund;
formation and regulation of the functioning of the grain product market.
Its implementation may include the following directions:
development of budget programs for stimulating of the agro-industrial production, food industry, sale of agro-food products, creation and development of market structures;
creation of a fund to support the agricultural producers;
providing the financing of priority areas of the development of agriculture in accordance to the target state programs;
development of the procedure of agricultural enterprises' debt restructuring;
ensuring the formation and development of the grain product market, the realization of its regulation, the expansion of its capacity attempting the entrance and consolidation in the foreign market.

Consequently, due to the generalization of the scientific material and data obtained from the poll of the respondents, the necessity of development of integrated formations in the grain product subcomplex is substantiated, the directions of the development of the grain market and the grain product subcomplex are summarized and grouped according to the degree of priority. Formation and development of integrated structures relocates the regional economic units into the system of competitive economic interrelations; provides strengthening of economic space of the territory by the means of innovation-oriented development in clusters, bringing production as close as possible to the sources of raw materials and places of consumption of final production, placing agricultural production evenly across the country, aimed at leveling the socio-economic development of the regions, strengthening the competitiveness of our country; and promotes the further development of the grain product subcomplex and its infrastructure.

References

1. Altukhov, A.I.: Russian Agroindustrial Complex: Current State And Problems of The Development. FGUP "VO Minselkhoz of Russia", 602 (2004)

2. Amin, A.: Living in the Global. Globalization, Institutions and Regional Development in Europe. Oxford University Press, pp. 1–22 (1994)
3. Araujo-Enciso, S.R., Fellmann, T.: Yield variability and harvest failures in russia, ukraine and kazakhstan and their possible impact on food security in the middle east and North Africa. J. Agri. Econom.T. **71**(2), 493–516 (2020)
4. Covey, T., Durst, R., Ryan, J.: U.S. Macroeconomic and tax policy: impacts on U.S. FArm operations. Agrekon (English Ed), T. **65**(2), 149–163 (2005)
5. Fedoseeva, S., Herrmann, R.: The price of sanctions: an empirical analysis of german export losses due to the russian agricultural ban. Canadian J. Agri. Econ. T. **67**(4), 417–431 (2019)
6. Hailu, G., Poon, K.: Do farm support programs reward production inefficiency? Canadian J. Agri. Econo. T. **65**(4), 567–589 (2017)
7. Kabbiri, R., Dora, M., Elepu, G., Gellynck, X.: A global perspective of food market integration: a review. Agrekon (English Ed), T. **55**(1–2), 62–80 (2021)
8. Odening, M., Shen, Z.: Challenges of insuring weather risk in agriculture. Agricultural Finance Review, T. **74**(2), 188–199 (2014)
9. Olipra, J.: Price transmission in (de)regulated agricultural markets. Agrekon (English Ed), T. **59**(4), 412–425 (2020)
10. Porter, M.E.: The Competitive Advantage of Nations. Macmillan, London (1990)
11. Porter, M.E.: On Competition. Harvard Business School Press (1998)
12. Rada, N., Liefert, W., Liefert, O.: Evaluating Agricultural Productivity and Policy in Russia. Journal of Agricultural Economics T **71**(1), 96–117 (2020)
13. Rungsuriyawiboon, S., Wang, X.: Investigating agricultural productivity improvements in transition economies. China Agricultural Economic Review, T. **4**(4), 450–467 (2012)
14. Yakovenko, N.A., Rodionova, I.A., Ivanenko, I.S., Kireeva, N.A., Sukhorukova, A.M.: Export potential as the competitiveness indicator of the agri-food complex. International Journal of Engineering and Technology T. **7**(4.38), 654 (2018)

Designing a Model of Cryptosecurity of Information in the System of Countering the Spread of COVID-19

Aleksandr Troshkov[1], Aleksandr Zhuk[2], Irina Kuzmenko[1](✉), Oleg Florinsky[3], and Valeriy Rachkov[2]

[1] Stavropol State Agrarian University, Stavropol, Russia
troshkov1954@mail.ru, 11kip11@mail.ru

[2] Stavropol State Agrarian University, North-Caucasus Federal University, Stavropol, Russia
alekszhuk@mail.ru, rw6hlg@mail.ru

[3] Pyatigorsk Institute (Branch) North Caucasian Federal University, Pyatigorsk, Russia
florinski@yandex.ru

Abstract. Information security of information systems is vital for the protection of information resources. If coronavirus 229E, NL63, OC43, HKU1 enters the human body, one of the therapeutic treatment tactics is used to restore the immune system. By analogy with this process, an equivalent of the cryptographic information protection model is proposed in this paper. The study of the attack of the COVID-19 virus and the therapeutic treatment of the human body made it possible to propose a cryptoprotection model based on the algorithm for combating the virus in question. The modern studies carried out by scientists from various countries, but most of all by Russian specialists, devoted to the adaptive immune response, were of decisive importance for the development of vaccines and preparations of monoclonal antibodies; on this basis, an algorithm for designing cryptographic protection of information that circulates in the system of transmission and reception through channels of various physical nature. Taking into account the equivalent of the therapeutic treatment of COVID-19, it is proposed to filter the user by biometric, characteristics or human parameters (BCP), (HBP) for the purpose of high-quality and prompt identification. For the scientific approval of the presented proposals, a mathematical confirmation based on the interpolation of the Lagrange polynomial is given.

Keywords: Crypto protection · Immune system · COVID-19 · Reaction · Equivalent · Information security · Dysfunctional response

1 Introduction

Innovation and technological progress only move forward with fair competition, but not all organizations and individuals are ready to compete fairly. This undermines the scientific potential and does not allow for accelerated development. The communication structure is developing rapidly. More than half of the world's population is connected to the Global Information Networks, almost the entire population of the world lives in the coverage areas of telecommunication networks. Digitalization in the innovation sphere,

I. Samoylenko (Ed.): AMBT 2022, LNNS 582, pp. 28–39, 2023.
https://doi.org/10.1007/978-3-031-20803-4_4

the qualification of specialists allows unscrupulous competitors to illegally obtain scientific or other information, which slows down investment policy. The COVID-19 crisis has increased digitalization remote activity, and at the same time, the number of attacks on information resources has increased and, as a result, information leakage in various directions. This can be countered by improving the quality of methods and ways of protecting information, including cryptographic algorithms. The adaptive immune response of COVID-19, taken as a basis, was crucial for the development of vaccines and monoclonal antibody preparations; on this basis, an algorithm for designing cryptographic protection of information that circulates in the transmission and reception system through channels of various physical nature is proposed.

A pandemic caused by the diagnosable respiratory disease Coronavirus-2 (SARS-CoV-2) caused by acute respiratory syndrome (SARS-CoV), which emerged against the backdrop of an exacerbation of a public illness. All the health sciences in immunology and infectious diseases have joined the fight against this persistent disease, which represents scientific developments in the fight against this virus. This article reviewed and considered the pathophysiology of SARS-CoV-2 infection and based on the presented algorithm for generating a model of cryptographic protection of potential and perceived information. Analyzing immunological research and detection gives us an understanding of the natural processes of host-pathogen interaction in COVID-19, which helps to reveal a picture of reliable information for determining the tactics of the study of the disease, including the discovery of new treatments [9]. In addition, the materials produced by modern research, specifically devoted to the immune response, are of particular importance for the development of vaccines and preparations of monoclonal antibodies. A flurry of information attacks fell upon humanity, as well as an increase in the number of thefts, destruction of integrity, fake information. It is proposed to take therapeutic effects on the SARS-CoV-2 virus as an equivalent and carry out information security modeling, which allows us to set the scientific goal of designing an algorithmic model of cryptographic information security [2]. It is fairly reliable and confirmed that coronaviruses cause disease in humans and animals. Four of these (human coronaviruses 229E, NL63, OC43, and HKU1) usually only cause upper respiratory tract infections and cause relatively mild symptoms [4]. Diseases in patients are caused not only by a viral infection, but also by the reaction of the host organism.

2 Materials and Methods

The purpose of the study is based on the possibility of using the cycle of the COVID-19 virus infection on the human body and the response of the human body and its immune system to the coronovirus infection to design an information security protection system. The objectives of the study were based on understanding the penetration of the COVID-19 viruses into the body and its effect on the defeat of human organs, as well as the concentration of the organism itself and the therapeutic effect of directed against viruses [5]. To fulfill the scientific task, theoretical methods were used, which consisted in systematizing the actual analyzed publication material, studying and testing theoretical provisions. This made it possible to combine the analyzed materials to obtain a different result on the formation of an information security algorithm that has novelty and relevance [10]. Based on the results of the selection of research methods, an empirical method was chosen, which made it possible to combine the analyzed materials to obtain a different result on the formation of an information security algorithm that has novelty and relevance. The materials of the pathophysiology of SARS-CoV-2 infection were analyzed, on the basis of which an algorithmic model of cryptographic protection of transmitted and received information was proposed. [3, 11, 12].

3 Results

Combination of processes of origin of entry into the body of the ARDS virus, observed in severe COVID-19, is characterized by shortness of breath and low levels of oxygen in the blood, and a powerful release of cytokines by the immune system in response to a viral infection leads to a cytokine storm and symptoms of sepsis, which is the cause of death in 28% of fatal cases of COVID-19, due to uncontrolled inflammation leading to multiple organ failure [6]. Such a process is widely represented in Fig. 1.

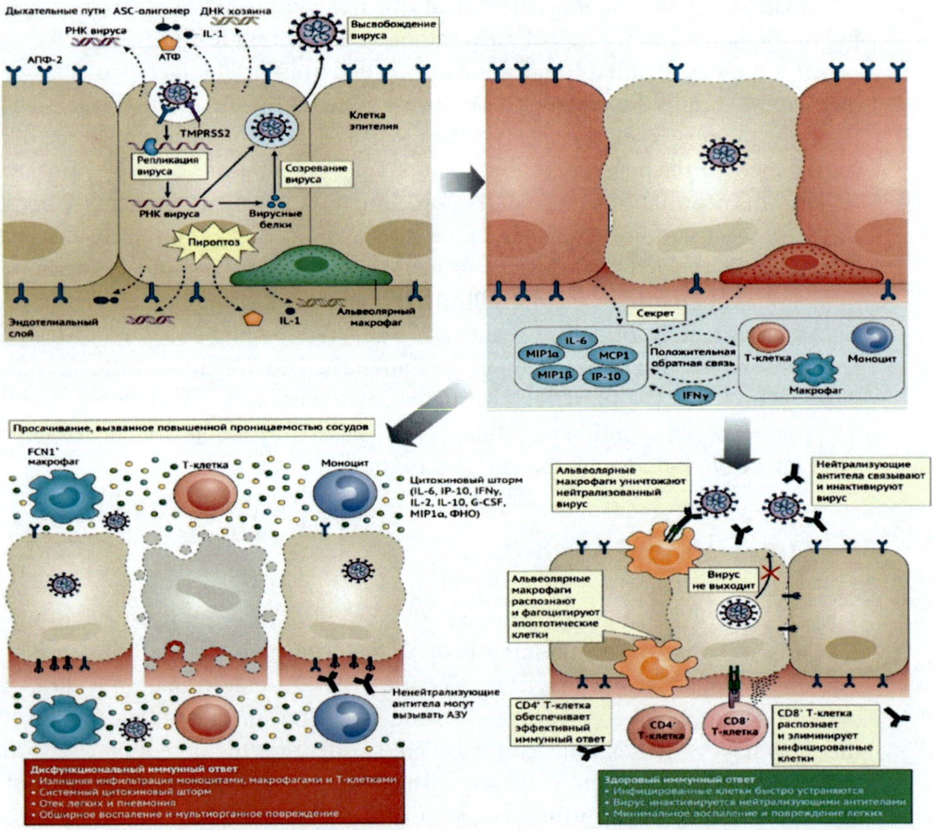

Fig. 1. Algorithmic sequence of COVID-19 cyclic activity

The description of Fig. 1 is proposed to be carried out in an algorithmic way using the algebraic operations presented in Fig. 2.

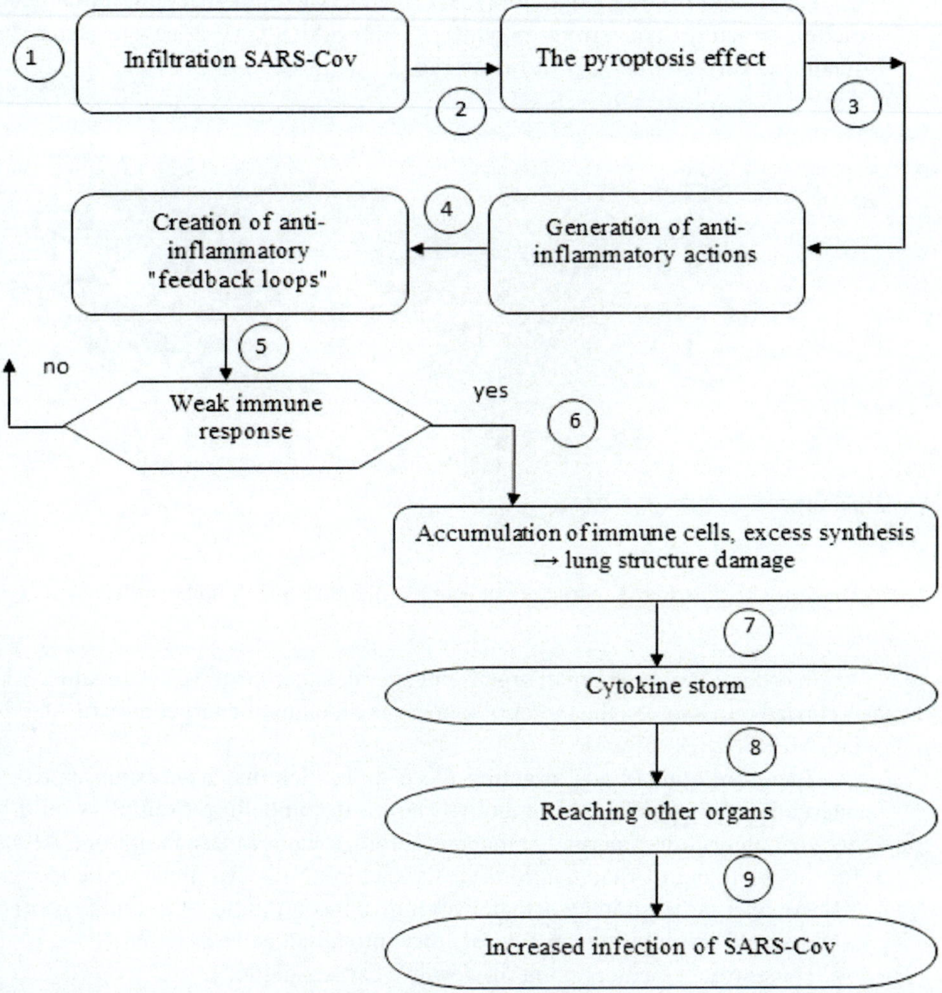

Fig. 2. Reflection of the results of the algorithm of algebraic operations of cyclic reality COVID-19

From Fig. 2 and analysis of the algorithm of algebraic operations of the cyclic activity of COVID-19, we can conclude that the main stage of infection is the binding of the virus to the host cell through the target receptor on epithelial cells of the respiratory tract, alveolar epithelial cells, endothelial cells of blood vessels. SARS-CoV infection reduces the expression of angiotensin-converting enzyme (ACE2) in lung cells. Because loss of ACE2 function in the lung is associated with acute lung injury, virus-induced suppression of ACE2. Medical studies have shown that ACE2 regulates the renin-angiotensin system (RAS) [7].

Therefore, a decrease in ACE2 function after infection with the virus may lead to RAS dysfunction affecting blood pressure and fluid and electrolyte balance, as well as increasing inflammation and vascular permeability in the airways.

The receptor-binding domain (RBD) interacts with its receptor, human angiotensin (action structure) converting enzyme (ACE2) SARS-CoV-2, severe acute respiratory syndrome coronavirus-2, is shown in Fig. 3 [1].

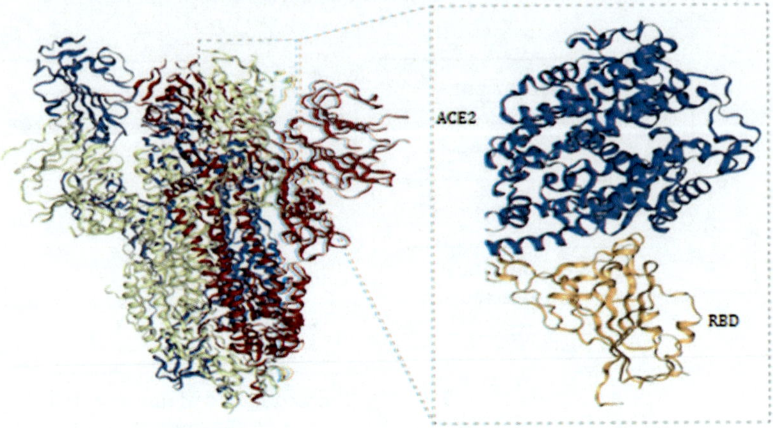

Fig. 3. Structure of action of the SARS-CoV spike protein

In order to save all human organs in general, a therapeutic effect against SARS-CoV is carried out, namely, the target of therapy is blocking the target human ACE2 receptor or TMPRSS2 [8] (Fig. 4).

There are compounds targeting these molecules that have been approved by the international therapeutic community. The most compelling example would be neural network algorithms that predict that baricitinib, a Janus kinase inhibitor (JAK) approved for the treatment of rheumatoid arthritis, can inhibit ACE2-mediated endocytosis [11, 13]. An alternative strategy is to deliver high concentrations of a soluble form of ACE2, which could potentially reduce viral entry into target cells.

Thus, from a medical point of view, we can conclude:

- antibodies (formed by vaccination or by adaptive transfer) block the interaction of SARS-CoV-2 with the ACE2 receptor in host cells;
- TMPRSS2 prevents cleavage of the spike protein, which is necessary for the introduction of the virus into the host cell;
- virus-specific CD8 + memory T-cells left over from a previous vaccination or infection can differentiate into effector cells during re-recruitment. When they identify infected cells presenting virus-specific epitopes, they degranulate and kill the infected cells before they can produce mature virions;
- in a new treatment that targets the symptoms of a cytokine storm, the blood of COVID-19 patients is passed through specialized reservoirs that are specifically designed to capture anti-inflammatory cytokines, after which the "purified" blood is returned to patients.

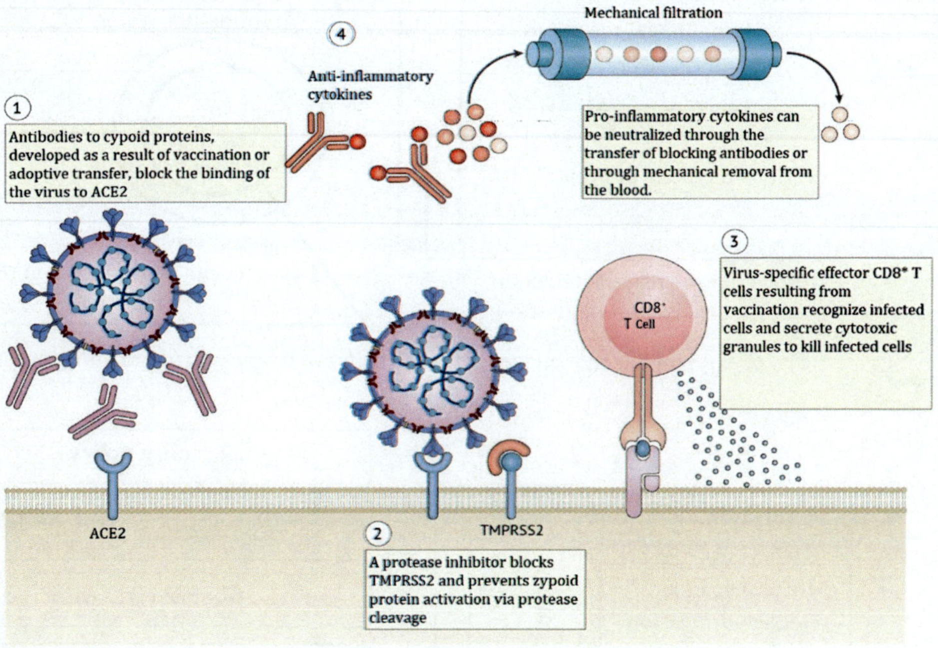

Fig. 4. The target of therapy is blocking the target host ACE2 receptor or TMPRSS2 (enzyme inhibitors)

4 Discussion

Based on the target of therapy, it is proposed to show a model of an attack on an information system (Fig. 5).

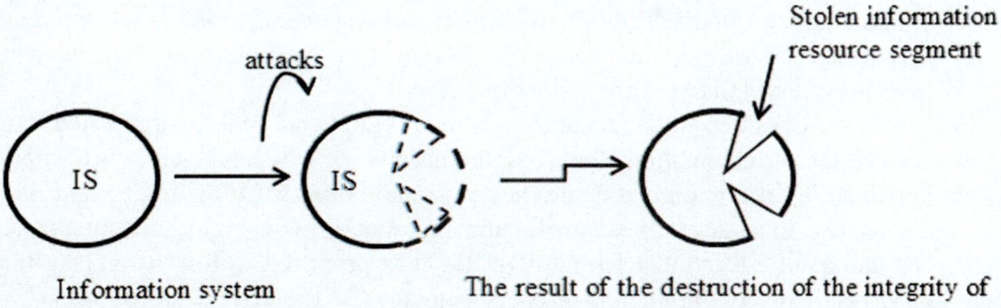

Fig. 5. An attack on an information system with the aim of destroying its integrity or destruction

Based on the model of attack on the information system (IS) and information resources (IR), it is proposed to consider the design model of crypto - security of IS and IR on the example of an information resource (Fig. 6).

To complement the model presented in Fig. 6, the main causes of unauthorized access to information and violations of technical information systems are presented, Fig. 7.

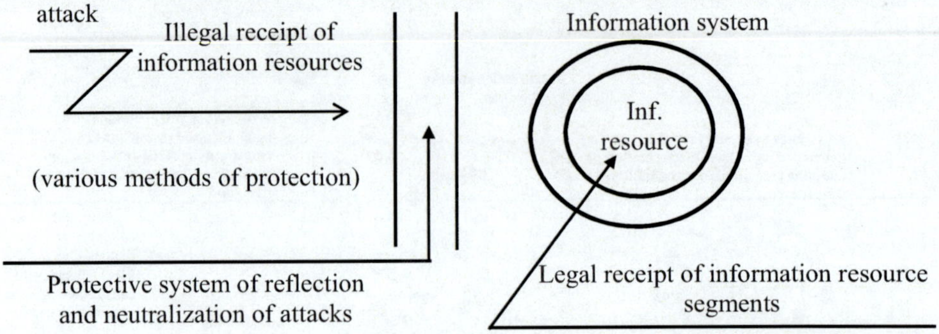

Fig. 6. Cryptoprotection model against unauthorized access to information resources

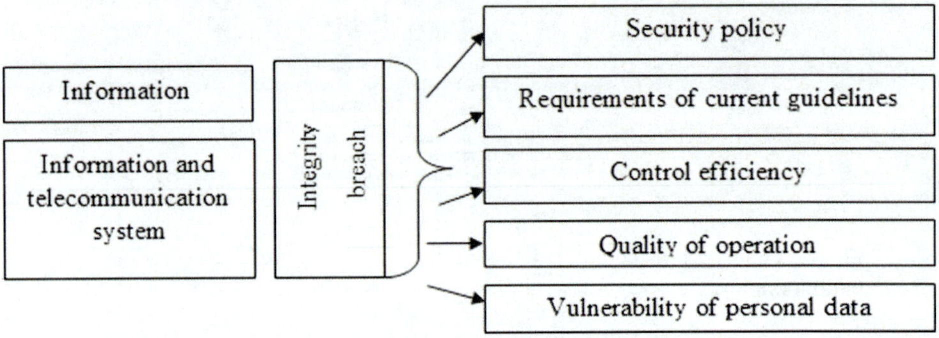

Fig. 7. Causes of unauthorized access

Analyzing the target of therapy - blocking the target ACE2 receptor of the host or TMPRSS2 and reflecting the results of the algorithm of algebraic operations of the cyclic reality of COVID-19, we can conclude that it is important to create an anti-inflammatory "feedback loop" (dysfunctional response), which is interpreted from the point of view of equivalence as the design of cryptoprotection from introducing the actions of cryptanalysts not "friendly" users.

If we consider and analyze Fig. 2, then digital encryption is proposed to apply the protection of information resources, the purpose of which is to close information while maintaining the main characteristics of signal-code structures (SCS) and having high resistance to attacks. However, at the first stage, taking into account the equivalent of therapeutic treatment for COVID-19, it is proposed to filter (user) by biometrics, characteristics or parameters of users (human) (BCP), (HBP).

The choice of biometric characteristics and parameters, which are carried out according to the following proposed algorithm (Fig. 8).

Based on the execution of the algorithm (Fig. 8), BCP and HBP are selected according to the presented method, taking into account the biometric agreement. Next, an analysis of the attack on the information resource or the transmitted carrier signal (information carrier) is carried out. The attack, as a rule, is deliberate, which means that the damage caused can be represented in the following interpretation (Fig. 9).

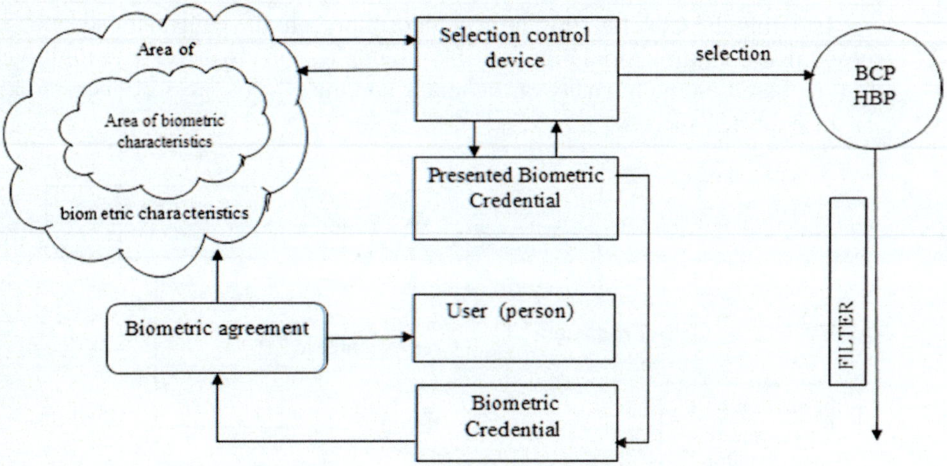

Fig. 8. Sampling algorithm for BHP and BBP

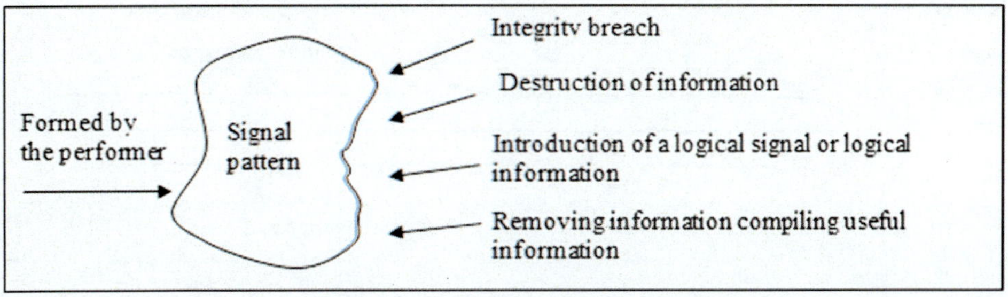

Fig. 9. Possibility of damage

Analyzing Fig. 9 and the equivalent of COVID-19 treatment, the creation of an "anti-inflammatory feedback loop" is proposed.

Based on possible violations and damage, it is necessary to identify the channel of attacks on information, that is, if considered with the previously presented equivalent, then this can be compared with the definition of a viral cell and its possible consequences in the human body. Therefore, a model is proposed that allows determining the share of cryptographic violations in the transmission of information, its processing and storage. Conducted judgments on this issue made it possible to propose a model for determining attacks and predicting damage from it (Fig. 10).

Thus, from the presented model (Fig. 10), you can achieve the following results: pre-determine the attack sites and the probability of losing the information part, as well as destruction of the integrity of the cipher message. For a more accurate assessment of the result, an expert level was used to determine the damage, which is assessed on a single quantitative scale.

The expert assessment level presented in Fig. 11 makes it possible, by equivalent - the generation of anti-inflammatory drugs, to form methods of encryption protective measures to prevent damage assessed on a single quantitative scale.

Isolating the COVID-19 treatment algorithm from the equivalent scheme, it is proposed to use multifactorial biometric cryptography, and the increase in multifactoriality will be based on the intensity of the attack and the effectiveness of decryption, as well as the expected success.

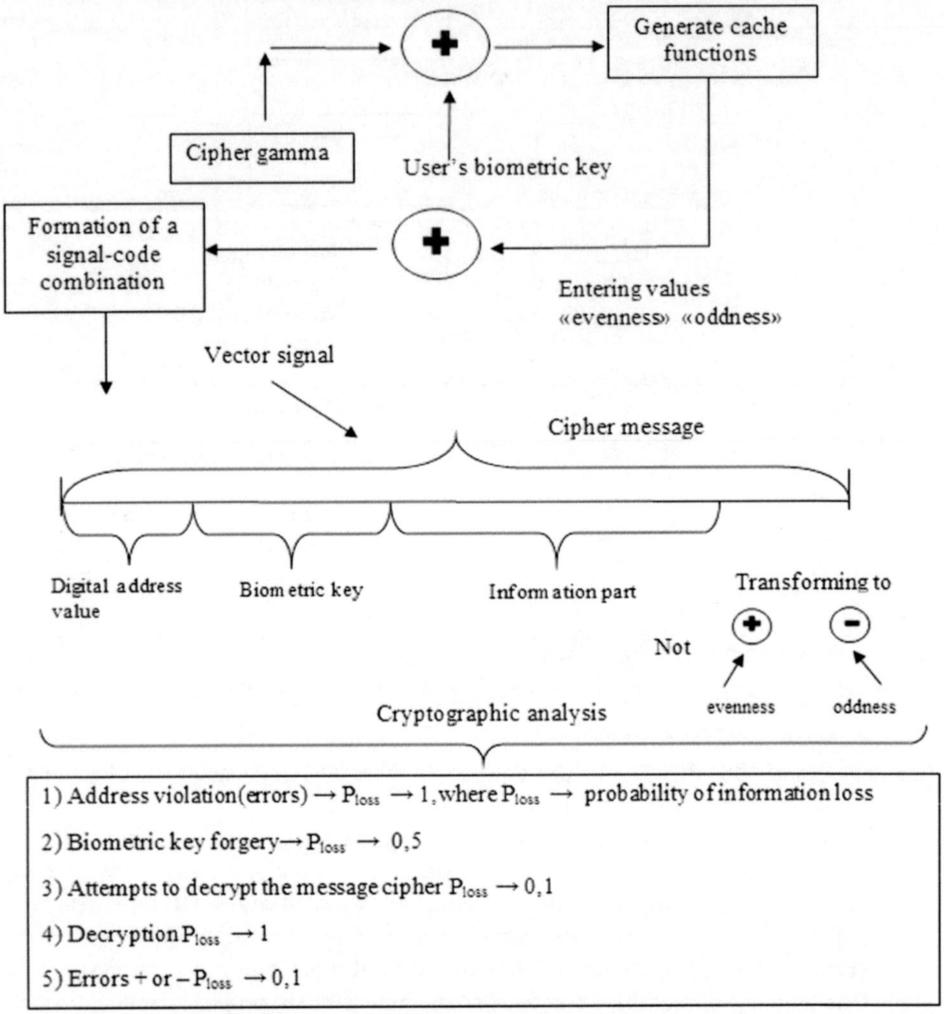

Fig. 10. Model for determining attack and damage

The success of this proposal lies in the fact that in building up multifactorial biometrics, a given set of BCPs (HBP) is used, consisting of simple BCPs to complex BCPs, and their integration increases the security of cryptographic information, which makes it possible to maneuver and perspective. However, there are limitations in the choice of biometrics in the set of presented and known BCPs, since due to the possibility of damage, it is necessary to complicate their application.

Figure 12 is an expanded block diagram of the therapeutic response equivalent.

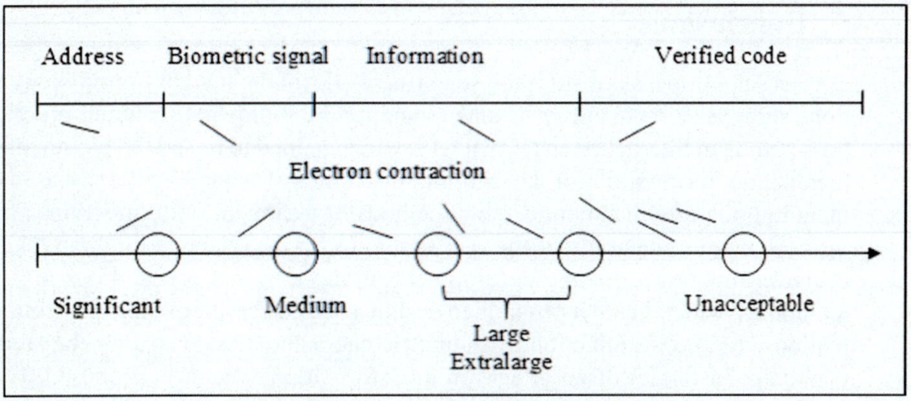

Fig. 11. Expert level assessment

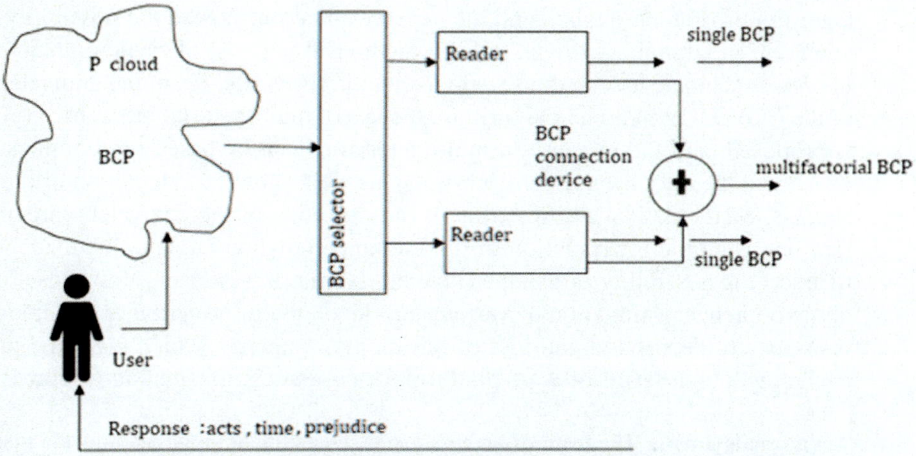

Fig. 12. Structural diagram equivalent to therapeutic response

Mathematical confirmation of the operation of the circuit (Fig. 11) can be carried out using the interpolation of the Lagrange polynomial (because layered BCG is used).

P is a cloud where the nature of the BHP and HBP consists of m-elements, since cryptographic sequences S are involved in the transmission, which consist of *j* parts carrying information of different secrecy *n*, then using a polynomial with the function *f(x)* we have:

$$\mathrm{F}\ (\mathrm{x})\ = a_0 + a_1 + a_2x^2 + \ldots + a_{j-1}x^{j-1} \tag{1}$$

then the segmentation (markedness) will be determined

$$\{p_i f\,(p_i)\} \tag{2}$$

From formulas (1) and (2) it is possible to confirm the possibility of using a (layered) biometric system as a response to cryptanalytic influence. Studies conducted on the model showed high resistance to decryption of protected information.

5 Conclusion

The set scientific task of studying scientific works in the field of immunology and infections showed the possibility of understanding the biological dynamic processes of the host-pathogen interaction in COVID-19, which helped to reveal the picture of important therapeutic information for determining the tactics of managing the diseased and them, including the establishment of new methods of treatment with infection. The analysis of the scientific materials of the recent studies on the adaptive immune response is of decisive importance for the development of vaccines and preparations of monoclonal antibodies, which made it possible to design a layered cryptographic protection of information with the introduction of biometric characteristics. Based on the studied target of therapy, a model of an attack on an information system is proposed. However, at the first stage, taking into account the equivalent of therapeutic treatment for COVID-19, it was proposed to filter (user) by biometric, characteristics or parameter of users (human) (HCh), (BHR). This made it possible to assume that the presented protection algorithm is of good quality, and the expert evaluation showed the possibility of generating anti-inflammatory drugs, to form methods (methods) of cryptographic protection to prevent damage assessed on a single quantitative scale. From the equivalent scheme of the COVID-19 treatment algorithm, the use of multifactorial biometric cryptography is proposed, and the increase in multifactoriality will be based on the intensity of the attack and the effectiveness of decryption, as well as on the calculation of the expected success. Mathematical confirmation of the operation of the equivalent circuit was carried out using the interpolation of the Lagrange polynomial, and this made it possible to assert the possibility of using a (layered) biometric system as a response to cryptanalytic influence. Simulation of various attacks on the information environment showed resistance to decryption and loss of information integrity, which gives the prospect of designing a high-level software platform for protecting information resources.

Acknowledgement. The team of authors expresses gratitude to the team of the medical community of the Stavropol Regional Clinical and Diagnostic Center for valuable consultations and clarifications on the clinical actions of COVID-19 and explanation of the infection counteraction processes.

References

1. Avila-Ponce, U. de Leon., Perez, A.G.K., Avila-Valez, E.: SEIARD's COVID-19 epidemic model in Mexico: mathematical analysis and state-level forecast. Chaos, solitons and fractals 140 (2020)
2. Wang, P., Zheng, S., Ai, G., Liu, D., Zhu, B.: Time series prediction of COVID-19 epidemic trends using advanced LSTM deep learning: case studies in Russia, Peru, and Iran. Chaos, Solitons and fractals 140 (2020)
3. Zhuk, A., Orel, D., Vanina, A., Minkina, T.: Information security threats of wireless smart utility metering systems. In: CEUR Workshop Proceedings this link is disabled **2842**, 143–150 (2021)
4. Koltsova, E.M., Kurkina, E.S., Vasetsky, A.M.: Mathematical modeling of the spread of the COVID-19 coronovirus in a number of countries in Europe, Asia, Israel and Russia, Problems of economics and legal practice. Moscow **N2**, 154–165 (2020)

5. Zhuk, A., Orel, D., Nekrasova, E., Krivolapova, O.: The use of simulation models and game scenarios in the study of radio engineering systems by higher engineering students. In: CEUR Workshop Proceedingsthis link is disabled, p. 2494 (2019)
6. Contreras, S., et al.: Multigroup SEIRA model of the spread of COVID-19 among heterogeneous populations. Chaos, solitons and fractals 136 (2020)
7. Komunian, A., Gaburro, R., Giudici, M.: SIR-based model inversion: critical analysis of application to the COVID-19 epidemic. Physica D. 413 (2020)
8. Kurkina, E.S., Koltsova, E.M.: Mathematical modeling of the spread of waves of the epidemic of the coronavirus COVID-19 in different countries of the world. Applied Mathematics and Informatics No. 66 - M. Publishing House of the Faculty of VMK MSU, pp. 41–66 (2021)
9. Kuzmenko, I., Chervyakov, N., Yuriy, K., Samoylenko, D.: Modification of the scheme of division of asmuth-bloom data with the application of the method of fractal geometry. In: Proceedings - 2017 4th International Conference on Engineering and Telecommunication, Enand T **4**, 28–32 (2017)
10. Kuzmenko, V.V., Glotova, I.I., Tomilina, E.P., Kuzmenko, I.P., Fedorenko, I.V.: Assessment Instruments of economic feasibility of tax transformations. Asian Social Science T. **11**(19), 298–305 (2015)
11. Marinov, T.T., Marinova, R.S.: Dynamics of COVID-19 using an inverse problem to identify coefficients in SIR epidemic models. Chaos, solitons and fractals, p. 5 (2020)
12. Best Practices for Infection Prevention and Control with a Focus on COVID-19: Countries Share Experiences. Epidemiology and Vaccinal Prevention. **1**, 76 (2020)
13. Tsaranov, K.N., Zhiltsov, V.A., Klimova, E.M., Tarbastaev, A.G.: Tech. perception of the threat to the personal safety of medical professionals in the context of the COVID-19 pandemic. Health Manager **4**, 15–19 (2020)

Check for updates

Ensuring Economic Security and Sustainable Development of the Southern Region of Russia

T. N. Polutina[1], K. E. Tyupakov[1(✉)], and V. K. Tyupakov[2]

[1] Kuban State Agrarian University named after I.T. Trubilin, Krasnodar, Russia
[2] Kutafin Moscow State Law University, Moscow, Russia

Abstract. In the context of global and regional instability, competition for sales markets, resources and cost effectiveness, political factors influence on economic processes, special attention is paid to economic security of Russia generally representing economic security of its regions. The objective of the work is to develop the target indicators for regional economic security monitoring, identify the positions and prospects of its development, a complex of measures aimed at its sustainable development. Based on the critical analysis of native and foreign scientists' works, laws and regulations, the authors provide the interpretation of the concept "regional economic security" represented as the state of socio-economic system ensuring the implementation of national interests and regional sustainable development. The mechanism of ensuring regional economic security is presented. 24 indicators for regional economic security evaluation are developed by an analytical method. Indicators characterize macroeconomic, innovative, social, environmental and foreign economic development of a region, as well as the level of its industrial, food, energy, financial and personnel security. The study is carried out for the southern Russian region - Krasnodar territory. The potential threats and risks in industry, energy, innovation and social development, environmental and foreign economic areas of the region are specified. A complex of measures of the regional economic security program is proposed. It assumes to increase investment and innovation activity in the real economic sector, attract the private capital to finance research and development, introduce the waste-free technologies, and improve the foreign trade activities regulation. The scientific novelty of the research is the clarification of a concept "regional economic security" and identification of analysis indicators and potential threats to Krasnodar territory economic security.

Keywords: Economic security · Region · Monitoring · Indicators · Threshold values · Threats

1 Introduction

Nowadays special attention is paid to Ensuring Russian economic security in the context of international economic penalties against the country, permanent underfunding of the real economic sector, technical and technological underdevelopment of the material and technical base and low innovation activity. Economic security is ensured both by early forecasting and identification of the threats to the national economic mechanism,

I. Samoylenko (Ed.): AMBT 2022, LNNS 582, pp. 40–52, 2023.
https://doi.org/10.1007/978-3-031-20803-4_5

and development of a complex of measures applied to prevent or localize the negative consequences of specified threats. The level of state, society and individual protection determines directly the stability of material, industrial, social, political and spiritual spheres of society.

Almost all world countries have adopted the relevant laws and regulations that are the basis for an integrated state policy formation in the field of military, economic, social, environmental, and information security.

There are a great number of approaches to the concept "economic security" in the scientific national and foreign literature. A lot of scientists associate economic security with the global economic system and its terms of reference – with unsteady socio-economic development, the external debt growth, the cyclical economic and food crises and other globalization aspects [4, 6, 13, 15].

Other researchers focus on the conditions and factors of national economy sustainable development, including free access to energy sources, liberalization of investments and opportunity of goods and technologies exchange [1, 3, 5, 16]. The International Red Cross Committee defines economic security as the capacity of individuals, households or communities to meet their basic needs.

Thus, the majority of implemented research is focused on economic security investigation at the global and national levels. In our opinion, national economy sustainable development is only possible by means of increasing of its competitiveness. In the context of structural transformation of world economy, increasing market volatility, instability of the international financial and economic system and reduction of real and financial investments flows it can be achieved by resistance of its regions, individual industries and enterprises against the negative impact of stochastic threats to economic security.

The objective of our article is to develop monitoring indicators, assess the level and justify the measures aimed at ensuring regional economic security and sustainable development.

2 Materials and Methods

The theoretical background of the research comprises the national and foreign economic scientists' scientific papers [1, 3–6, 9, 10, 13–16], laws and regulations of the Russian Federation and its constituent entities [2, 8, 12], statistical data of socio-economic development of Russia and its regions [7, 11], existing techniques for the level of economic security evaluation. The research is developed on the results of a comparative evaluation of socio-economic development of Russia and its regions based on calculation of the indicator average value for the period of 2015–2020. The initial statistical information has become Rosstat and Krasnodarstat data.

Based on the calculation results, a system of indicators characterizing the security level of all spheres of the region's economy has been formed. The indicators threshold values have been defined as well. The threshold values of the share of imported goods in industrial and food potential, national expenditures on research and development, educational funds, health and social policy, the share of persons engaged in scientific research were defined in accordance with international comparisons and expert conclusions.

When specifying the indicators, it has been considered that the threshold values are determined by consideration period, economy development, geographical and natural features. The number of indicators in the system is sufficient to implement all necessary evaluations. The degree of indicator deviation from the threshold value and its dynamics characterize threat realization probability. Achievement of the threshold values of socio-economic development indicators determines stability of the economic, environmental and social processes of a region and its capacity for self-development and progress.

Based on the research results, the economic security level and its threats applied to Krasnodar territory have been determined and a complex of measures to strengthen economic security has been presented.

3 Indicators Characterizing Regional Economic Security

The actual Russian law, the concept of "security" is defined as "protection of individual, society and state vital interests from internal and external threats". It first appeared in the Law of the Russian Federation "On Security" as of 05.03.1992, which gave the grounding for the national security system. Personality, personality's rights and freedoms, society, its material and spiritual values, state, its constitutional system, sovereignty and territorial integrity were defined as the main security objects [8].

The National Security Strategy of the Russian Federation was approved by the Decree of the President of the Russian Federation No 400 as of July 2, 2021, in accordance with the federal laws "On Security" and "On Strategic Planning in the Russian Federation". The named strategy defines national interests, aims, tasks and ways of national security ensuring. The concept "economic security" is interpreted as "protection of national economy from external and internal threats which ensures economic sovereignty of the country, unity of its economic area, the conditions for implementation of the strategic national priorities of the Russian Federation" [2].

Thus, economic security is characterized by ownership relations, economic mechanism and development of the institutions of state power providing sustainable and balanced development of national economy. Economic security management is aimed at countering the risks and threats to the country's economic system. It is provided through implementation of a complex of measures oriented to maintenance of its stability and economic potential development.

A special role in ensuring national economic security is assigned to the particular region capabilities to form and develop the state and regional economic potential. The regional economic system is a complex of relations between economic entities striving to sustainable growth. It is characterized by the specific production features creating the differences in the combination of conditions and factors, institutions and mechanisms aimed at protecting the economic interests of a region and improving its residents' lives.

The mechanism of regional economic security provision is presented in the Fig. 1 and includes the elements of managerial relations, organizational and economic instruments, the measures to prevent, neutralize and eliminate the threats, as well as the economic crimes. In other words, regional economic security is the state of socio-economic system providing implementation of the national interests and regional sustainable development.

The key background for regional economic security provision is implementation of the activities of strategic regional program comprising the laws and regulations that codify the threats to national and regional economic interests, the counteraction measures and effectiveness control of managerial decisions. The object of regional economic security is all areas of real and financial-credit economic sector, consumer markets and trade, households and the individual.

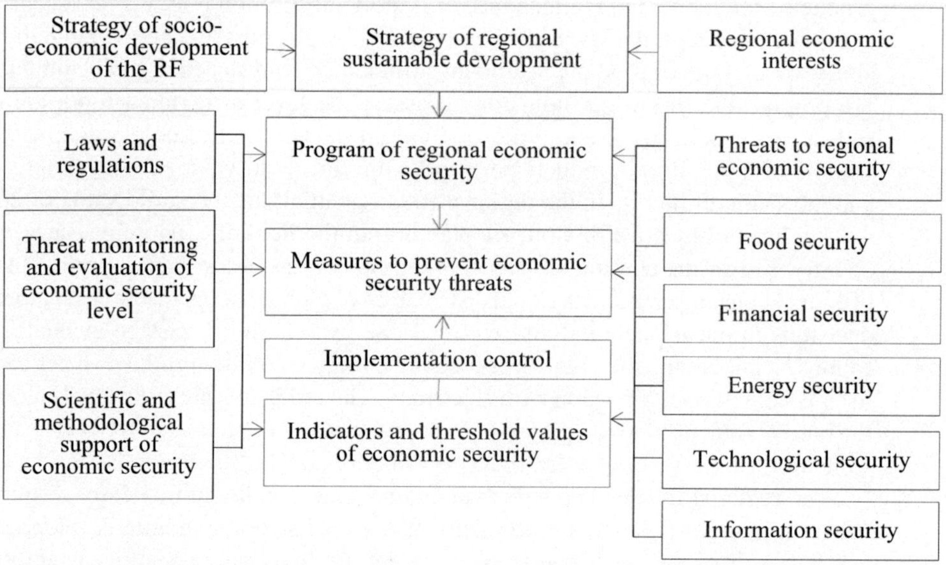

Fig. 1. Mechanism of ensuring regional economic security

Ensuring economic security is impossible without its level evaluation and threats monitoring. Economic security monitoring is an integral part of its mechanism and should guarantee accuracy and timeliness of the values obtained. The monitoring is based on differentiation of the indicators of socio-economic development grouped into the integrated blocks characterizing macroeconomic, innovative, social, environmental and foreign economic development, as well as the level of industrial, food, energy, financial and personnel security.

Macroeconomic development of a region is characterized by the indicators of per capita gross regional product and the annual inflation and unemployment rate. The indicators threshold values are fixed at the national average level and recalculated subject to the deflator indexes.

An important indicator characterizing regional economic security is the level of industrial development. The share of industry in Russian economy is over 30% and its development is one of the key focuses of socio-economic strategy. The state policy implemented in the industrial sphere is aimed at elimination of the endemic problems slowing down innovative development of its industries.

It can be highlighted the following problems: low labor productivity, high energy and material consumption, a high level of depreciation of fixed assets, lack of financial opportunities and other factors. We propose using three indicators to evaluate region industrial security: the level of depreciation of fixed industrial assets, the threshold value of 40%, the volume of per capita industrial production, the threshold value of

448.9 thousand RUB and the share of imports in industrial potential, the threshold value of 20%.

A high growth rate of national economy and its constituent entities is impossible without agricultural sector development achieved by the innovative solutions for agricultural products production and maintenance of import substitution policy. Food security, both at the state and regional levels, is characterized by physical and economic public availability of all types of food in the amount sufficient to lead an active and healthy lifestyle. We propose the following indicators to assess the level of regional food: the level of self-dependence in the basic types of food products, the threshold value of 90%, the volume of agricultural products per capita, the threshold value of 44.6 thousand RUB and the share of imports in the regional food potential, the threshold value of 20%.

The major objective of strategic planning in the field of economic security of each country is to ensure continuous access to the energy resources at a fair price. The growth of energy consumption and unequal distribution of global energy resources require timely decisions to ensure national and regional energy security. It can be evaluated by the following indicators: the volume of fuel and energy resources production per capita and the level of regional provision with electricity. The indicator threshold level calculated as the ratio of total production per country population is 21.3 thousand RUB. The threshold value of the balance of electric energy production and consumption is 1.

The regional financial sphere ensures financial, credit and monetary resources distribution in order to solve the problems of regional socio-economic development. The key indicator of this system effectiveness and its functioning security is zero balance of regional consolidated budget. The balance between revenues and expenditures of regional consolidated budget is achieved by a high level of economic development and the growth of population's real disposable incomes. The indicator of own funds share in the revenues of consolidated budget is of particular importance. The share of taxes on property, personal income, profits, and uncompensated receipts should be at least 75% of the total regional revenue, otherwise, it may lose its financial sovereignty.

We consider personnel security through entrepreneurial activity efficiency due to the regional workforce potential growth. It can be evaluated by the following indicators: the number of students enrolled in the higher and secondary vocational education programs and the number of employees at the scientific institutions. The threshold value is determined by the method of international comparisons. It is 300 people per 10 thousand of population and 2.5% of the number of people engaged in the economy. Development of a network of specialized secondary, higher and scientific institutions, a variety of educational programs and practice-oriented training will provide the regional economy with highly qualified human resources.

Innovative development plays a decisive role among the economic security indicators, since the efficiency level of socio-economic system in general depends on timeliness and the quality of innovations introduced in the regional economy. The higher are the research and development costs, the higher are the share of innovative products in the economic sectors. The internal research and development costs are assessed in % of regional gross product, and the threshold level established on the basis of international comparisons is 2.2%. The share of innovative products in regional industry should be at least 25%.

Regional social development is characterized by the following indicators: the ratio of average per capita income to minimum subsistence level and the amount of funds for education, health and social policy. Since the subsistence minimum level is calculated as a cost of a conditional consumption basket and provides a minimum standard of living, the population's incomes (wages, business income, interests on deposits, social and other official incomes) must exceed it at least 3 times for an acceptable standard of living. It is proposed to fix the threshold value of the level of funds aimed at the social sphere at 12% of the revenues of regional consolidated budget.

Regional ecological state is closely related to social development. Living standards and conditions depend on the volume of harmful substances in the environment, therefore, the threat to environmental security appears if discharge of contaminated waste water and emission of pollutants into the atmosphere exceeds the established value of 5.0 m3 per km^2 and 6.0 thousand tons per km^2.

In the context of world trade decline, foreign economic development is a key priority in the national and regional socio-economic development strategy and regions. Russian foreign trade turnover is over $ 550 billion USD, and its 80% come from Moscow and St. Petersburg, Moscow, Leningrad, Kaliningrad, Sverdlovsk, Kaluga regions, Primorskii and Krasnodar territories and the Republic of Tatarstan. Exports plays an important role in ensuring economic security, so, we propose to assess the level of foreign economic development of a region by a coefficient of import–export coverage ratio and the volume of exports per capita with thresholds of 1 and 20.000 USD, respectively.

Based on regional economic security evaluation, a relevant program is developed, which includes the following implementation priorities:

1. to overcome the consequences of the global and national economic crisis, achieve economic growth in regional economy real sector and subject it to the objectives of national and regional socio-economic development.
2. to achieve food security and self-dependence by increasing the efficiency of regional agro-industrial complex and import substitution.
3. to enhance regional financial security by means of monetary, credit, banking and fiscal policy.
4. to increase the level of regional energy security through energy conservation policy and development of regional energy potential, diversification of electricity sales markets and creation of conditions for competition in the field of energy provision.
5. to ensure technological safety by means of updating and upgrading the enterprises and organizations technical and technological base.
6. to improve environmental security by preventing extreme environmental pressure.

4 Analysis of the Economic Security Indicators of Krasnodar Territory

Krasnodar territory is one of the most important Russian regions and a leading one in the Southern Federal District. It possesses a well-developed educational system, large scientific and production potential, developed industrial, construction, fuel and energy complexes, the sphere of information and communication technologies, as well as agro-industrial, transport, resort, recreational and tourist complexes. The Strategy of socio-economic development of Krasnodar territory was adopted in 2018 and corresponds to Russian priorities of socio-economic development and determine a special status of Krasnodar territory in the national economy [12].

Krasnodar territory covers the area of 75.5 thousand km^2 in the southern part of the Russian Federation, in the south-west of the North Caucasus. More than 5.5 million people live in the region, about 53% in the cities and 47% in the rural areas. The average population density is 66.6 people per 1 km^2. The largest cities are Krasnodar (948.8 thousand people), Sochi (432 thousand people), Novorossiisk (340 thousand people) and Armavir (187.2 thousand people).

Dynamic development of Krasnodar territory economy is primarily subject to the richest resources of agricultural land, including chernozems, whose area is 4805 thousand hectares, which is 4% of all-Russian and 2% of world reserves. Krasnodar territory ranks first among the constituent entities of the Russian Federation in terms of production of grain, sugar beet, fruits and berries. It is the second in production of sunflower seeds and honey, eggs, livestock and poultry meat (in live weight) and the third in milk and vegetables production.

The ratio of Krasnodar territory in the structure of some all-Russian indicators is shown in the Fig. 2.

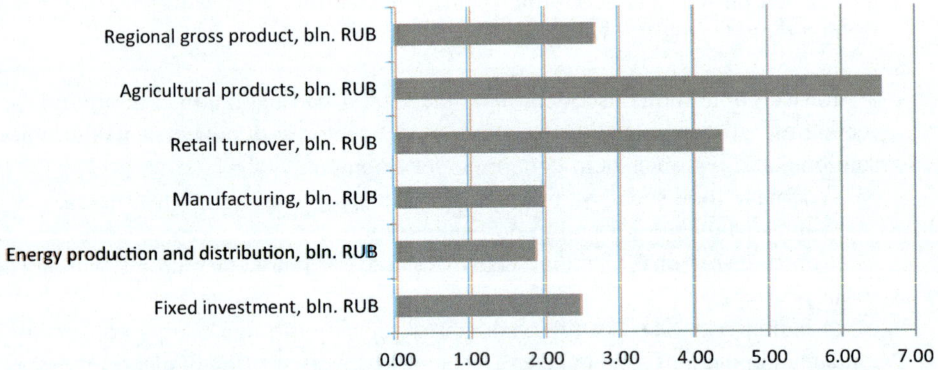

Fig. 2. Ratio of Krasnodar territory in all-Russian indicators (2020)

Retail trade is one of the key consumer sectors in Krasnodar territory. The variety of trade formats with stable consumer demand is motivation for the medium and small businesses. Oil refining industry, chemical and petrochemical production, and energy industry are developed by large and medium-sized enterprises producing petroleum products, electricity, mineral fertilizers, sulfuric acid, plastic products, fiberglass, paint and rubber products. Krasnodar territory demonstrates the high rates of investment activity. The total volume of investments has an annual growth rate of 5–6% and exceeded 319 billion RUB by 2021.

We have developed a system of indicators to evaluate the level of economic security of Krasnodar territory. It includes 24 indicators grouped into 10 blocks reflecting macroeconomic, innovative, social, environmental and foreign economic development, as well as industrial, food, energy, financial, personnel security of the region. The Table 1 demonstrates the indicators of economic security of the region.

Table 1. Economic security indicators of Krasnodar territory

Indicators	2020 г	Threshold values
Regional gross product per capita, thous. RUB	502.6	516.0
Annual inflation rate, %	4.2	5.0
Unemployment rate, %	4.2	4.0
Level of depreciation of fixed industrial assets, %	37.2	40.0
Industrial output per capita, thous. RUB	191.6	448.9
Share of imported goods in industrial potential, %	24.2	20.0
Self-dependence in the basic types of food products, %	93.6	90.0
Volume of agricultural products per capita, thous. RUB	70.4	44.6
Share of imported goods in food potential, %	35.2	20.0
Extraction of fuel and energy resources per capita, thous. RUB	11.0	21.3
Ratio of electricity generation to consumption	0.38	1.0
Balance of regional consolidated budget to regional gross product, %	−0.04	0.0
Share of own funds in consolidated budget revenues, %	79.2	75.0
Ratio of public debt to own income, %	30.0	30.0
Number of students of secondary and higher education institutions per 10,000 persons	390.3	300.0
Number of individuals engaged in research activities to employed persons, %	0.25	2.5
Internal expenses on research and development to regional gross product, %	0.7	2.2
Share of industrial innovation, %	2.2	25
Ratio of average per capita income to minimum subsistence level, %	3.1	3.0
Funds for education, health and social policy, %	7.6	12.0
Discharge of polluted waste water, m^3 per region area, km^2	8.6	5.0
Emissions of pollutants into the atmosphere by stationary sources, thous. tn. per 1 km^2	6.3	6.0
Import–export coverage ratio	1.25	1.0
Volume of exports per capita, thous. RUB	1.1	2.0

Macroeconomic Development
The regional domestic product of Krasnodar territory has increased by 48% over the past 5 years and amounted to 2890 billion RUB by 2020. Its cost per capita exceeded 500 thousand RUB and amounted to 97.4% of the national level. The annual inflation rate was 4.2%, which is below the threshold one. The number of unemployed people is 160 thousand people. The unemployment rate in the considered period decreased by 0.7 percentage points and did not exceed the threshold value of 4%. It should be noted that the coefficient of tension in the region is 0.6%, i.e. there is a sufficient number of jobs for most of the citizens looking for employment.

Industrial Security
The industrial production accounts for about 17% of the regional gross product. It produces over 590 thousand tons of oil, 1670 million m3 of gas, 500 thousand tons of clay, 4.5 million m3 of sand and other minerals. The food production is valued at about 400 billion RUB, coke and oil production – 158 billion RUB, rolled metal and various metal products – at 80 billion RUB.

The industrial production volume reaches 1 trillion RUB, and the average number of industrial workers exceeds 200 thousand people. The average annual value of industrial fixed assets is 1 trillion RUB. Their physical and moral depreciation reached 37.2% with a threshold value of 40%. The industrial production volume includes the cost of produced and sold gas, water and electricity, goods of manufacturing industries, as well as extracted minerals per capita of the region. This indicator is fixed at the level of 448.9 thousand RUB per capita, while this indicator was only 191.6 thousand RUB or 42% of the threshold value in Krasnodar territory. The share of machinery, equipment, vehicles, consumables, components and finished products of imported production in the region is 24.2%. It exceeds the threshold value and poses a threat to the regional industrial security.

Food Security
Agricultural production of Krasnodar territory increased by 17.3% during the considered period and reached 400 billion rubles in 2020, which is almost 2 times higher than the threshold value per capita. The crop production is more than 68% of the manufactured products and increased its volume by 8.2%. The livestock production increased by 43% and reached 126 billion RUB in value terms.

Self-dependence in the main food products, considered as the ratio of regional production and domestic consumption in Krasnodar territory, reaches 94%, which exceeds the established threshold. The share of imported products in the food potential of Krasnodar territory is quite high and amounts to 35.2%, which is 15.2 percentage points higher than the threshold value. A large share in imports is occupied by vegetable products – 36%, fats and oils – 10%, food products, including beverages and tobacco – 6.1%, animal products – 1.7% of the total import volume.

Energy Security
Krasnodar territory is the oldest oil-producing Russian region. There are oil and gas deposits in the Azov-Kuban cavity. Annually, the cost of extracted fuel and energy resources is about 10 billion RUB are extracted in the region. Electricity generation in

the region is provided by 40 thermal and 2 hydroelectric power plants with the total capacity of 2,370 MW. However, the energy security analysis has shown that the region does not have any significant energy resources reserves. It is an energy-deficient region, and the lack of electricity is compensated by the flows from the neighboring power systems. The cost of extracting fuel and energy resources is almost 2 times lower than the threshold value, and the share of electricity generated in the region does not exceed 38%.

Financial Security
Krasnodar territory ranks second among Russian regions in terms of public debt (100 billion RUB). The budget loans comprise its major part. The threshold value was reached by a high share of own funds in the consolidated budget revenues, as well as the consolidated budget balance. The last one was ranged from -0.9 to 1.2% to the regional gross product in the examined period. The financial security indicators have reached the threshold value, what characterizes a high level of financial security in the region.

HR Security
The number of students enrolled in the secondary and higher education programs is 220 thousand people. It is 390 people per 10 thousand of population and higher than the threshold value. However, the number of people engaged in science does not exceed 7000 people, which is 0.25% of the number of people employed in the region at the level of threshold value of 2.5%.

Innovation Development
Currently, over 6.5 billion RUB are directed to the regional scientific and educational institutions for research and development. It is 0.7% in relation to the regional gross product. The share of innovative products in Krasnodar territory does not exceed 2.2%, with a threshold value of 25%.

Social Development
The of Krasnodar territory population's actual incomes amount to 35.6 thousand RUB. It is 3.1% in relation to the subsistence minimum which is higher than the accepted threshold value. The expenditures on education, healthcare and social services account for 7.6% of Krasnodar territory consolidated budget, which is 4.4 percentage points lower than the threshold value.

Ecological Development
Waste water discharge into the water bodies of Krasnodar territory is carried out by the enterprises of heat-power, food, oil refining and mechanical engineering industries. Its annual volume is about 3000 million m3, which forms 67% of the total volume of waste water in the Southern Federal District. Pollutants emissions in the region amount to 6.3 tons per km2, and 90% of them are from the stationary sources. In terms of environmental pollution, Krasnodar territory is included in the TOP-10 Russian regions along with

Krasnoiarsk territory, Khanty-Mansiisk autonomous district and Primorskii territory. The indicators of environmental pollution exceed the threshold values by 5–70%.

Foreign Economic Development

Krasnodar territory trade turnover with the foreign countries is 10.2 billion USD with the share of far-abroad countries of 90%. The balance of foreign trade turnover was positive in the amount of 1116 million USD. The import commodity structure is dominated by food products, chemical products, machinery and equipment. The mineral products, agricultural raw materials, metals and their derivatives prevail in the commodity structure of exports. The exports amounted to 5638 million USD with the current imports volume at the level of 4521 million USD, what made it possible to obtain an import–export coverage ratio of 1.25. The exports volume per capita in Krasnodar territory amounted to 1.1 thousand USD with a threshold value of 2 thousand USD.

It order to prevent and eliminate the threats to Krasnodar territory economic security which were identified during the analysis, we have created a complex of measures aimed at regulation of its socio-economic system:

1. Investments attraction in manufacturing industries development (mechanical engineering and metal working, building materials production, chemical, glass, wood working and light industries). It will allow the enterprises and organizations to implement timely modernization and technical re-equipment of production, expand production capacity, reduce the depreciation level of fixed assets and increase the production volume.
2. Private capital attraction to provide research and development with financial support subject to the concession agreements. It will create the research structures at enterprises and organizations, increase production efficiency and reduce the share of imported goods in the regional production potential.
3. Development of a consolidating regulatory legal act in the field of innovation. Its application will permit to consider the goals of all stakeholders and participants of innovation process (the state, a municipality, organizations, business incubators, technology parks, etc.), stimulate the regional innovative activities and increase the share of innovative products in the regional economic sectors.
4. Introduction of a complex of methods to reduce harmful substances emission in the environment at industrial enterprises and introduction of waste-free, low-waste and resource-saving technologies in agro-industrial complex. It will increase the protection level of natural objects, human life, health and ownership interests.
5. Introduction of the international quality standards at the regional export-oriented enterprises what will increase national products competitiveness in the world market.

5 Discussion and Conclusion

The research has established that the level of regional economic security plays an important role in socio-economic development of Russia. The considered theoretical and applied aspects of ensuring economic security have made it possible to develop a system of indicators for assessing the regional economic security. The system characterizes macroeconomic, innovative, social, environmental and foreign economic development of the region, as well as the level of industrial, food, energy, financial and personnel security. The presented target indicators fully specify the opportunity of achieving the regional economic security and allow implementing its quantitative and qualitative evaluation.

The evaluation of Krasnodar territory economic security has detected the threats to sustainable development of industry, energy, innovation, social, environmental and foreign economic spheres. The crisis situations have occurred due to the disturbing factors, such as output reduction at the industrial companies, domestic goods displacement from the market, destruction of production and technical potential, economy deindustrialization, scientific research reduction, disappearance of the organizations engaged in scientific research, personnel loss in the field of complex technologies, problems of regional financing, instability of financial resources directed to the social sphere, growth of harmful emissions, increase of man-made loads and natural systems imbalance. Export increase of raw materials instead of the goods produced using advanced technologies reduces its quality and foreign exchange earnings, puts negative pressure on the foreign trade balance and slows down the processes of technological renewal.

The developed anti-crisis measures are aimed primarily at increasing investment and innovation activity in the real economic sector, attracting the private capital to finance research and development, introducing the technologies that reduce harmful emissions into the environment and improving foreign trade activities regulation. The research practical significance is the use of developed methodology of regional economic security evaluation and the complex of measures aimed at strengthening its socio-economic system by managers and specialists of legislative and executive authorities. It will help them to develop regional economic security programs and socio-economic development strategies.

We see the further research prospects in the investigation of legal-regulatory and scientific-methodological support of the regional economic security. Only legal-regulatory acts allow establishing the rights and obligations of all participants of economic activity, and the scientifically based methods of threat assessment help to obtain timely and reliable results.

References

1. Abalkin, L.I.: Economic security of Russia: the threats and their repulsion. Journal of economic issues, No **12**, 4–16 (1994)
2. Decree of the President of the Russian Federation: On the National Security Strategy of the Russian Federation (No 400). Printing House of the Government of Russia, Moscow, Russia (2021)
3. Senchagov, V.K.: Economic security of Russia. M. Delo (2005)
4. Andruseac, G.: CES Working Papers Series. Economic Security – New Approaches. In: The Context Of Globalization, pp. 232–240. Iasi, Romania. CES Alexandru Ioan Cuza University (2015)
5. Glazev, S.: Basics of ensuring national economic security – alternative reformation course. Russian economic journal, No **1**, 3–19 (1997)
6. Kahler, M.: Economic security in an era of globalization: definition and provision. Journal of Pacific Review **17**, 485–502 (2004)

7. Department of the Federal State Statistics Service for the Krasnodar Territory and the Republic of Adygea: Statistics yearbook: Statistics collection. KrasnodarStat, Krasnodar. Russia (2020)
8. The act of the Russian Federation: On security. (No. 2446–1). Printing House of the Government of Russia. Moscow, Russia (1992)
9. Mitiakov, S.N., Mitiakov, E.S.: Analysis of the crisis phenomena in Russian economy with the use of the rapid indicators of economic security. Journal of Forecasting problems **3**(186), 29–40 (2021)
10. Mitiakov, E.S.: Methodology and tools development for economic security monitoring in Russian regions. Thesis for a Doctor's of Economics Degree. Printing House of the Volga Region State University of Technology, Ioshkar Ola, Russian (2019)
11. Department of the Federal State Statistics: Russia in numbers. Brief statistics collection: M. Russia, Rosstat (2020)
12. The Law of the Krasnodar Territory: Strategy of socio-economic development of Krasnodar territory. Printing House of the Krasnodar Territory administration. Russia, Krasnodar (2018)
13. Tamoshiuniene, R., Munteanu, C.: 1st International Conference on Business Management. Current research approaches to economic security. Printing House of the Management Polytechnic University of Valencia, Valencia, Spain (2015)
14. Tiupakov, K.E., Saifetdinova, N.R.: Economic evaluation of agricultural production technological development. Journal of Agroindustrial complex: economics, management **12**, 72–77 (2011)
15. United Nations General Assembly: International economic security. 119th plenary meeting, 40 session. Printing House UN, NY (1985)
16. Vechkanov, G.: Economic security: Course book for the HEIs. St. Petersburg, Russian: Piter (2007)

Financial Sustainability of an Agricultural Organization as a Condition for Its Economic Development

Lyubov Vinnichek(✉) and Dina Badmaeva

Saint-Petersburg State Agrarian University, Saint Petersburg, Russia
l_vinnichek@mail.ru

Abstract. The process of economic activity of the organization in modern conditions is subject to the continuous impact of a large number of external and internal factors that can significantly hinder its development and put in risk the possibility of economic growth. A complex characteristic of the success of the organization's business processes and management efficiency is financial sustainability. In the theory and practice of financial management general approaches to the analysis of financial stability, based on the study of the structure of capital of the organization, identifying the degree of financial independence of the business and the provision of reserves by own sources are recommended. Agricultural production has historically developed a capital structure with a predominant share of own sources and a high share of inventories in current assets, resulting in an inherently high degree of financial stability of the organization. But at the same time in economic activities can be fixed ups and downs in sales, net profit, return on capital. This in no way reflects the sustainability of the organization's development, its progressive economic growth. In the scientific study the methodology of analysis of financial sustainability of agricultural organization was proposed, the stages of which are based on the analysis of creditworthiness and evaluation of the effectiveness of agricultural production. The system of indicators of financial sustainability of agricultural organization is proposed for realization of the analytical task. Particular attention in this system is paid to the analysis of material and biological liquidity ratios, the monetary content of net profit and the impact of economic value added on the increase in equity. The implementation of the proposed methodology in practice is designed to provide a real assessment of the relationship between the financial stability of the organization and its economic growth, contribute to the validity of management decisions being developed.

Keywords: Sustainability of development · Sustainability of production · Financial stability · Solvency · Indicators of financial sustainability

1 Introduction

Under modern conditions of economic and business development, the problem of ensuring the sustainable development of the economic entity in the agricultural sector becomes

I. Samoylenko (Ed.): AMBT 2022, LNNS 582, pp. 53–63, 2023.
https://doi.org/10.1007/978-3-031-20803-4_6

particularly relevant. Successful progressive development of agricultural production serves as a basis for sustainable development of the whole national economic complex of the country [12].

Agricultural production is a sector subject to the negative impact of various factors of climatic, political, social, epidemiological nature, the manifestation of which causes unprofitability of activity and hinders the development of enterprises [2].

The problem of sustainability in the economic system of society has been worrying scientific minds for decades. The theory of sustainable development, which appeared in the second half of the twentieth century, questioned the unlimited nature of the possibilities of increasing production and consumption in connection with the limited natural resources on the planet and the increasing damage to the environment [7]. At the same time, the current growth of agriculture has been achieved mainly in an extensive way [8], so the issue of sustainable development of agricultural organization requires an urgent review and solution.

Sustainable agricultural development can be achieved by ensuring the reproduction of productive capacity, human resources and the natural environment for a longer period of time [6].

The general stability of an organization presupposes such a cash flow movement, which ensures a constant excess of income over expenses. Financial sustainability is a kind of mirror of such excess stably formed at the enterprise. The analysis of financial sustainability is based on the study of ratios, reflecting the structure of capital of the organization. In this case, the higher the degree of financial risk of activity, the lower the financial sustainability of the organization and vice versa. In our opinion, this conclusion is not always economically correct and justified.

For an agricultural organization it is advisable to evaluate financial sustainability on the basis of determining the organization's ability to uninterruptedly finance material and biological current assets, maintain the efficiency of production and reproduction of capital, ensure the growth of the monetary content of net profit. The purpose of scientific research is to develop a methodology for analyzing the financial sustainability of an agricultural organization.

2 Materials and Methods

The development of modern economic operations in the activities of business entities is accompanied by a high degree of uncertainty, the presence of crisis phenomena, tightening of the market environment for the functioning of organizations [1]. In these difficult conditions, the sustainability of agricultural organization development depends on timely and quality management decisions in all spheres of activity. Particular attention should be paid to the problems of the organization of financial flows, the efficiency of movement of which is the key to the financial sustainability of the organization.

Concerning the definition of economic essence of financial sustainability there are many points of view, having either common features or significant differences. Some authors interpret financial sustainability as the stability of the financial position of the enterprise, provided by a sufficient share of equity [10]; others define financial sustainability as the ability to assess financial capacity to withstand various external and internal influences and fulfill all obligations [5]; others under financial sustainability understand the state of financial resources, their distribution and use, which provide, under acceptable risk, uninterrupted operation, sufficient [13]. Most authors present financial sustainability in a narrower way than it should be, expressing financial sustainability in two or three indicators.

Financial sustainability is the ability of the enterprise to maintain a normal financial condition under adverse influences of factors of internal and external environment through the optimal structure of capital and assets, the competent ratio between assets and sources of their formation, the effective use of resources and rational reinvestment policy [5].

The object of the scientific study is the economic activity of the Joint Stock Company Breeding Plant Krasnogvardeyskiy of the Leningrad region of the Russian Federation. The main specialization of the organization is dairy cattle breeding. JSC Breeding Plant Krasnogvardeyskiy is one of the leaders among the agricultural enterprises of the district. As the research shows, in recent years the farm has been systematically working steadily to increase production volumes, which allows it to achieve high financial performance and actively introduce advanced technologies and equipment.

Financial sustainability of the agricultural organization is a complex characteristic of assets and capital, formed as a result of the interaction of diverse material and financial movements, capable of generating economic benefits in the coming period and provide as a result of this increase in equity. In this regard, scientific research is built on a comparative analysis of the dynamics of changes in capital, assets, profits and cash flows of the organization.

In accordance with the objectives of the study used methods of summary and grouping of economic data, comparison and dynamics, financial ratios, graphic. The study was carried out on the basis of the accounting statements of the organization. Analytical calculations were carried out using the software product Microsoft-Excel.

One of the generalizing indicators of financial sustainability and economic growth of the organization can be considered an increase in the value of net assets. In JSC Breeding Plant Krasnogvardeyskiy the net assets for the period from 2015 to 2020 increased more than twice (Fig. 1).

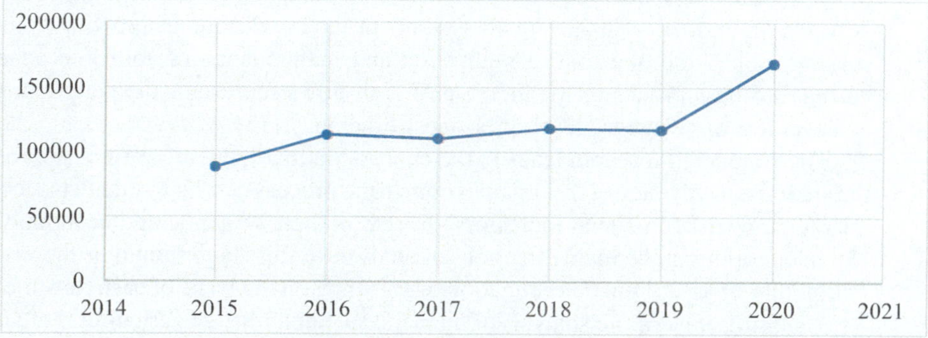

Fig. 1. Dynamics of increase in net assets of the organization, thous.

Judging by the figure, in JSC Breeding Plant Krasnogvardeyskiy for the past six years there has been an economic growth of activity, which allowed to provide an increase in net assets for the period by 716 million rubles.

In the study it was proposed to allocate three stages of analysis of financial sustainability of the agricultural organization (Fig. 2).

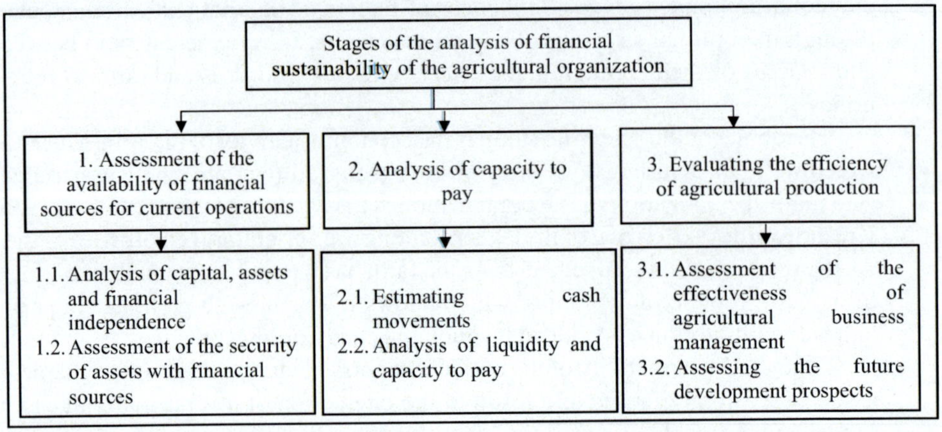

Fig. 2. Description of the analysis of financial sustainability of the agricultural organization

The analysis of financial sustainability of agricultural organization begins with the study of the composition and structure of capital. Due to objective laws of technological and biological nature, agricultural production requires the use of capital in the form of material objects and biological organisms. This determines the presence in the composition of the assets of the agricultural organization of a larger share of fixed assets, tangible and biological current assets, financed mainly from its own financial sources. Therefore, in the process of analysis it is necessary to determine the degree of security of the most important types of assets by financial sources.

At the second stage of the financial sustainability analysis it is necessary to evaluate the state of solvency, which is an external manifestation of the financial sustainability of the organization. Solvency analysis should be based on methodical approach, which is based on cash flow management system. In the conditions of growing competition for any subject the question of realization and maintenance of ability to generate by current business processes the increase of cash flows represents the paramount task of achievement of necessary level of business efficiency [13].

The conducted research leads to the conclusion that there are various approaches to the analysis of solvency of the organization, in the process of which scientists recommend a large number of financial indicators, diverse in their semantic and economic content [3]. In our opinion, the main directions of analysis at this stage should be the assessment of liquidity of assets and determination of solvency on the basis of cash flow movement.

The third stage of the analysis of financial sustainability of agricultural organization is devoted to the study of the efficiency of agricultural production. Modern agricultural organization is a complex system consisting of many interrelated and interacting elements, the assessment of the effectiveness of which can include a variety of indicators.

Financial sustainability of agricultural organization development largely depends on the quality of management, i.e. on the effective organization of the financial management system of the enterprise.

The implementation of the proposed stages of the analysis of financial sustainability involves the calculation and assessment of the following system of financial indicators (Table 1).

Table 1. The system of indicators of financial sustainability of agricultural organization

Indicator	Calculation methodology	Symbols
1. Assessment of the availability of financial sources for current operations		
Equity share	$R_{FI} = \frac{EC}{A}$, (1)	R_{FI} – financial independence ratio; EC – equity capital; A – assets
Availability of MCA and BCA's own funds	$R_{MBCA} = \frac{NWC}{MCA+BCA}$, (2)	R_{MBCA} – availability ratio of MCA and BCA; MCA – material current assets; BCA - biological current assets
2. Solvency analysis		
Material liquidity ratio	$R_{ML} = \frac{MCA}{CL}$, (3)	R_{ML} – material liquidity ratio; MCA - material current assets; CL – current liabilities
Biological liquidity ratio	$R_{BL} = \frac{BCA}{CL}$ (4)	R_{BL} – biological liquidity ratio; BCA – biological current assets
Credit solvency ratio	$R_{CS} = \frac{CCI}{CO}$ (5)	R_{CS} – credit solvency ratio; CCI – Current cash income; CO – credit obligations
Net income cash content ratio	$R_{NICC} = \frac{NCFCO}{NP}$ (6)	NCFCO – Net cash flow from current operations; NP – net profit
3. Evaluating the efficiency of agricultural production		
Capital productivity index	$I_{CP} = \frac{IA}{C}$ (7)	I_{CP} – capital productivity index; IA – income amount; C – average annual capital
Profitability of equity capital	$P_{EC} = \frac{NP+Am}{AC}$ (8)	P_{EC} – Profitability of equity capital; NP - net profit; Am - amortization; AC – average annual equity capital
Share of economic value added in net asset accretion	$S_{EVA} = \frac{EVA}{\Delta NA}$ (9)	EVA – economic value added; S_{EVA} – share of EVA in net asset accretion; ΔNA – net asset additions

The proposed system of indicators allows performing an integral assessment of the financial sustainability of the agricultural organization, which makes it possible to present a complete picture of stability (or instability) of its development.

A required condition for ensuring a smooth production process and maintaining the continuity of capital turnover of the agricultural organization is the constant availability of material current assets in the form of seeds, feed, petroleum products, mineral fertilizers, as well as biological current assets in the form of young animals and animals in growing and fattening.

The ability of an agricultural organization to meet these current needs from its own sources is paramount to the financial sustainability of the entity.

3 Results

The formation of a certain level of financial sustainability of the agricultural organization occurs in the process of production, processing and sale of agricultural products, works and services to buyers and customers.

In the course of scientific research various absolute and relative indicators, most often recommended for the study in the analysis of financial sustainability of commercial organizations, were analyzed and investigated. This made it possible to give the author's definition of the term under study in relation to an agricultural enterprise.

Financial sustainability of agricultural organization - an integral characteristic of its financial condition in the environment, expressed through the following components: maintaining the optimal structure of capital and assets, maintaining solvency and the ability to continue normal business activities, management of financial risk and return on capital to maintain and increase the market value of the business.

Let's analyze the dynamics of the provision of current activities in JSC Breeding Plant Krasnogvardeyskiy with financial sources (Table 2).

Table 2. Dynamics of current activity security ratios, units.

Indicators	2015	2016	2017	2018	2019	2020
Equity share	0,74	0,72	0,79	0,83	0,87	0,89
Availability of MCA and BCA's own funds	1,27	1,38	1,17	1,16	1,36	1,38

During the study period, the organization steadily increasing share of its own sources, which indicates a high degree of ability to finance the need for material and biological current assets - this indicates the use of the farm rational and science-based farming systems, feed production and animal husbandry.

The availability of an abundance of the most important types of current resources allows the farm to maintain a high degree of liquidity and control the ability to meet credit obligations (Fig. 3).

The values of biological and material liquidity ratios ranged from 2.5–3.0 in 2016 to 10–12.5 in 2017. The high level of these ratios in 2017 was due to a sharp decrease in the

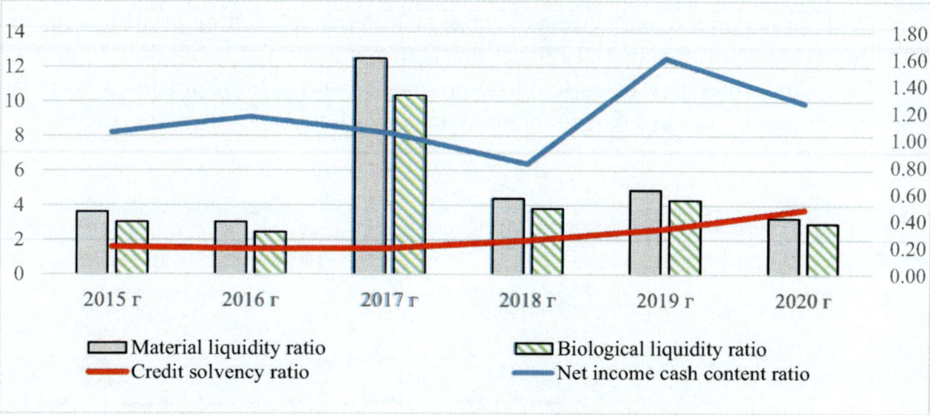

Fig. 3. Analysis of solvency ratios of an agricultural organization

amount of current liabilities. In general, the values of material and biological liquidity ratios are quite stable and testify to a high degree of liquidity of the current assets of the organization.

In JSC Breeding Plant Krasnogvardeyskiy long-term financial sources in the form of bank and leasing loans prevail in the composition of borrowed funds. As of the end of 2015 the organization was able to repay 20% of all credit liabilities at the expense of monthly incoming cash on current activities; as of the end of 2020. - already 48%. This indicates the improvement of settlement and payment relations of the farm with financial creditors.

Under market conditions, the goal of entrepreneurial activity is to make and generate profit. At this stage of development, when the external economic climate of business has significantly worsened for various political, environmental and social reasons, the monetary content of the net profit received by the subject is of paramount importance. In JSC Breeding Plant Krasnogvardeyskiy, the lowest value of this ratio was recorded in 2018, when net income of 120.3 million rubles was supported by cash receipts of 98.6 million rubles, i.e. by 82%. The best indicator of this coefficient is recorded in 2019. (160%), in this year the farm achieved a net profit of 90 million rubles, and cash inflows from current activities amounted to 144 million rubles.

At the next stage of the analysis we will study the evaluation of the efficiency of agricultural production in JSC Breeding Plant Krasnogvardeyskiy. Production efficiency can be expressed as an indicator that reflects the ability of an economic entity to produce the maximum amount of products with the minimum use of resources and the existing level of technology [4].

As mentioned above, the efficiency of production in the agricultural organization depends on a number of factors, the fundamental ones are: in crop production - the organization of crop rotation system, reclamation work; in animal husbandry - the system of breeding, maintenance and feeding of animals. At the level of the farm as a whole the most important factor is the system of organization and management of economic activity, motivation of management personnel and the entire staff of the organization.

With an integrated and systematic approach to the accounting of all factors, competent and purposeful managerial impact on their manifestation is possible to ensure the stability and efficiency of agricultural production and achieve financial sustainability of development. Let us analyze the recommended financial indicators (Fig. 4).

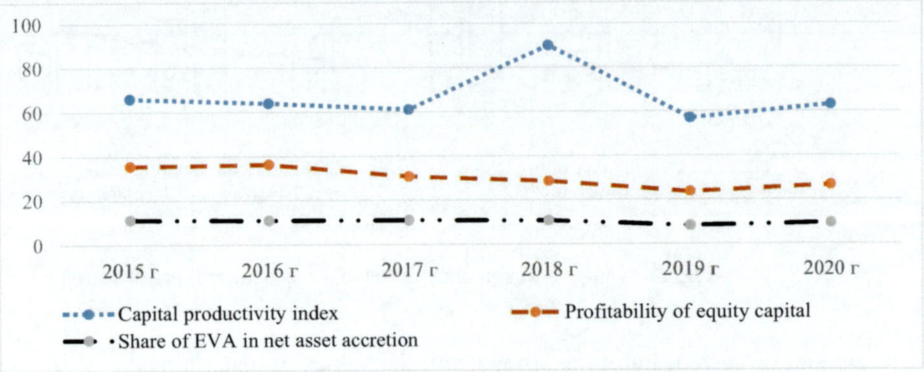

Fig. 4. Analysis of agricultural production efficiency indicators, %

As shown in the figure, the farm's level of capital productivity is stable, with the exception of 2018. The sharp increase in capital productivity in this year was due to an increase in other income in the activities of the organization associated with the disposal of land and fixed production assets. The growth rate of total revenues in 2018, compared to the previous year, was 64%.

Profitability is considered as a characteristic of efficiency [9]. The indicator of profitability of the equity is offered to analyze, taking into account the sum of the accrued for a year depreciation of fixed assets. As research has shown, at the expense of use of depreciation the profitability of the equity rises from 63 to 124% a year. As a negative moment we note a decrease in the return on equity for the period by 25%.

The amount of economic value added is a net increase in the value of the subject's equity created solely by the quality of the organization and management of economic activity. The economic value added is calculated according to the following methodology:

$$EVA = (R_{IC} - WACC) \times C, \tag{10}$$

где EVA – economic value added; R_{IC} – return on invested capital; WACC – weighted average cost of capital; C – average annual capital.

The dynamics of the EVA is shown in Fig. 5.

A comparative analysis of the changes in the economic value added indicator and the increase in net assets allows us to speak of a direct relationship between these indicators, which indicates the sustainable financial development of JSC Breeding Plant Krasnogvardeyskiy. Each ruble of increasing net assets generates on average about 10 pennies of economic value added in the organization, which, in our opinion, is a fairly high indicator of stability and sustainability of economic growth for an agricultural enterprise.

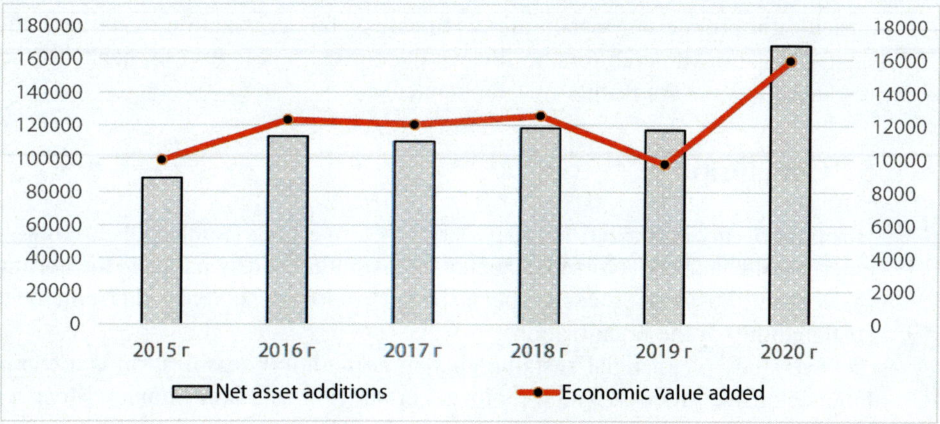

Fig. 5. Comparative analysis of performance indicators, thous.

4 Discussion

Financial sustainability is a multidimensional characteristic not only of the financial state of the organization, but also an integral criterion of the effectiveness of its economic growth. The stronger is financial sustainability, the higher is the possibility to provide economic growth of the entity.

The study of the system of financial sustainability indicators in JSC Breeding Plant Krasnogvardeyskiy allows us to make a significant increase in the scale of economic activity of the organization.

Agricultural production in the farm is organized in such a way that allows to achieve an annual gross profit, taking into account budgetary support, on average up to 115 million rubles. Stable profitability of economic activity allows the organization to provide the build-up of equity capital, the share of which has increased for the period by 20.5%.

To assess the impact of financial sustainability on the economic growth of the organization in the research study developed and proposed a methodology for analyzing financial sustainability, which includes analytical steps and calculation of the system of interrelated financial performance indicators.

As a result of the study it was found that in some years in the JSC Breeding Plant Krasnogvardeyskiy there were processes, the course of which can not be interpreted in a single way, in particular: a sudden decrease in the volume of short-term debt financing, a sharp increase in the amount of other income, a significant increase in cash receipts for current activities. This led to a jump in economic indicators of financial sustainability. The revealed tendency of financial indicators is not, in our opinion, the basis to draw a conclusion about violation of stability of development of the organization. In the activity of any economic entity there are always different situations and phenomena, the course of which may have a positive or negative impact on the change of indicators.

A distinctive feature of the proposed approach to the analysis of financial sustainability can be called its complexity and focus on the study of the cash flow of capital in agricultural production, which allows to assess the organization's ability to meet payment obligations to creditors, to finance the current and long-term needs for necessary resources and ensure further development.

5 Conclusion

Business of any organization today is exposed to a huge number of different, diverse, external and internal factors that can not only simultaneously paralyze the normal course of economic business processes, but also lead to a loss of solvency [11], and the financial sustainability of the organization.

The study of financial sustainability of agricultural organization is a complex and labor-intensive process, which requires certain time and labor inputs. Most scientists-researchers in the analysis of financial sustainability of an organization pay attention to the financial nature of this category, sincerely believing that the presence in the capital of the organization of a larger share of own sources and their constant increase act as a precondition and condition of financial sustainability and, ultimately, the sustainability of the economic entity.

It is worth recognizing that to some extent this statement is justified. Orientation in development of economic activity on own means of financing gives some advantages in the form of the minimum financial risks and absence of constant financial tension in settlement-payment relations of the organization. But at the same time such financial "freedom" essentially limits possibilities of realization of new capital investments, introduction of the innovations connected with cultivation of new varieties of plants, farm animals [6], production technologies that can hamper effective development of the organization.

Sustainable development of an agricultural organization is a dynamic process of quantitative and qualitative changes in the use of resources, achieving and maintaining an optimal scale of activity to increase economic value added.

The use of the proposed methodology of financial sustainability allows you to give a clear and distinct picture of the sustainability of the development of the agricultural organization. In the future the research of problems of financial sustainability of agricultural organizations can be directed to the development of normative criteria of financial indicators and the construction on this basis of the scale of financial sustainability of the organization.

References

1. Altukhov, A.L., Silaeva, L.P.: Improvement placement as a factor of sustainable development in agriculture. Studies in Systems, Decision and Control. T. **283**, 427–433 (2021)
2. Badmaeva, D.G.: Analysis of the solvency and sustainability of the development of agricultural organizations. Izvestiya MAAO **54**, 85–92 (2021)
3. Badmaeva, D.G.: Management of liquidity and solvency of agricultural organizations: Monograph. Moscow, Rusajns
4. Bershitskiy, Y.I., Saifetdinova, N.R., Tyupakov, K.E., Vasilieva, N.K., Artemova, E.I.: Justification of the structure and size of major productive resources of agricultural organizations. Int. J. Appli. Bus. Eco. Res. **14**(10), 7267–7287 (2016)

5. Bukhonova, S.M., Doroshenko, Yu.A., Benderskaya, O.B.: Complex methodology for analyzing the financial stability of an enterprise. Economic analysis: theory and practice **7**(22), 8–15 (2004)
6. Chaplitskaya, A., Heijman, W., Van Ophem, J., Kusakina, O.: Innovation policy and sustainable regional development in agriculture: a case study of the Stavropol Territory Russia. Устойчивое развитие. **13**(6), 3509 (2021)
7. Efimova, O.V., Volkov, M.A., Koroleva, D.A.: Analyzing the Impact of ESG Principles on Asset Returns: An Empirical Study. Finance: theory and practice **25**(4), 82–97 (2021)
8. Samygin, D.Yu., Baryshnikov, N.G., Mizyurkina, L.A.: Models of scenario forecasting development of agriculture in the region. Economy of the region. T. **15**(3), 865–879 (2019)
9. Samygin, D., Imyarekov, S.M., Tolmacheva, N.P., Kargin, Y.: Strategic management of modern financial requirements in agriculture. IOP Conference Series: Earth and Environmental Sciencethis link is disabled **341**(1), 012214 (2019)
10. Takhumova, O., Ryakhovsky, D., Satsuk, T., Kochkarov, P., Mirgorodskaya, O.: Ways to assess and improve the financial sustainability of Russian organizations development. Int. J. Eng. Adv. Technol. **9**(1), 1568–1571 (2019)
11. Vinnichek, L., Badmaeva, D.: Analytical tools for assessing the liquidity of assets of agricultural enterprise. Leasecon 2021: Conference on land economy and rural studies essentials, pp. 847–852 (2021)
12. Vinnichek, L., Badmaeva, D., Akhmadov, R.: Conceptual approach to the management of current assets of an agricultural enterprise. SHS Web of Conferences **110**, 04013 (2021)
13. McKenzie, W.: Use and interpretation of financial statements. Balance Business Books, Dnepropetrovsk (2006)

Digitalization as a Priority Direction of Russian Agricultural Development

Yuri Bershitskii, Aleksandr Sayfetdinov(✉), and Polina Sayfetdinova

Kuban State Agrarian University named after I.T. Trubilin, Krasnodar, Russia
sajfetdinov.a@kubsau.ru, polina.puzeychuk@bk.ru

Abstract. The article classifies the existing and promising digital technologies in agriculture in terms of complexity of applied equipment, information and communication and computer software. The factors possessing the greatest impact on effectiveness of agricultural management system digital transformation are specified and systematized. The digitalization analysis of technological process management system in the agricultural sector of Russia and Krasnodar territory are implemented. Additionally, the indicators of expected investment efficiency and riskiness in the process of development of innovative precision agriculture technologies with digitalization elements by the regional agricultural organizations are determined. The importance and effectiveness of widespread introduction of precision crop and livestock farming technologies into national agriculture are grounded and proved. Moreover, the land use boundaries for the agricultural producers of Krasnodar territory are established. The investments in development of precision crop farming technology lower than the fixed level are economically impractical and high-risk.

Keywords: Digitalization · Precision agriculture · Technical and technological agricultural innovations · Investment efficiency and riskiness

1 Introduction

Nowadays, one of the most significant challenges facing global agriculture is the deterioration of production conditions related to the world climate changes and population growth. These factors provoke the lack of food, what can be overcome only by using of various industry innovations. Thus, according to the forecasts available, at breakthrough growth of food demand, the planet population will exceed 9.5 billion people by 2050. Productivity of land and water production resources will tend to decrease.

One of the priorities of global agriculture innovative development is technological processes digitalization implemented mainly on the basis of widespread development of precision crop and livestock farming technologies [2, 6]. The share of agricultural producers using these advanced production technologies is 60–80% in the countries with developed economies, while it is only about 5% in Russia [3]. Thus, the accelerated technical and technological modernization in Russian agriculture based on development of precision crop and livestock farming technologies with the elements of digitalization

I. Samoylenko (Ed.): AMBT 2022, LNNS 582, pp. 64–72, 2023.
https://doi.org/10.1007/978-3-031-20803-4_7

is the most important direction of increasing industry efficiency which requires deep economic justification.

The research has identified and systematized the factors determining effectiveness of Russian agriculture digitalization. The existing and promising digital technologies in agro-industrial complex have been classified, the level of digital infrastructure development in the rural areas has been determined, and investment effectiveness and riskiness in innovative projects implementation in the field of agriculture digitalization has been calculated.

2 The Concept of Digitalization and Classification of Digital Technologies in Agriculture

Agriculture started actively developing and applying the digital technologies in the late 1980s, when the systems of machine units global positioning and the methods of remote earth sounding became widespread. It met well the system of conducting agro-industrial production with its seasonality, dependence on weather conditions and large areas of land. Currently, the task of agriculture digitalization is reduced mainly to the usage of big data in digital form for making management decisions based on artificial intelligence [8].

We consider national agriculture digitalization as deep technical and technological industry modernization with creation of a group of specialized digital platforms to increase labor efficiency and productivity in the agro-industrial complex (AIC), as well as timely information support when making managerial decisions at the different levels [4, 7].

All of this involves collection and digitization of production and economic information in automatic mode at all stages of final products value creation in AIC. Inclusion of end-users data in the information flows, who can also receive the necessary information about applied technologies used and products quality before sale. It allows the producers to adjust production solutions in a timely manner in order to satisfy the current consumer preferences as much as possible.

The next industrial revolution is based on the systems of electronic management of decision-making using big data, the Internet of things, artificial intelligence and machine learning. Presently, these digital technologies are introduced in agriculture in the form of elements of precision crop and livestock farming. The machinery, equipment and living biological objects are combined based on the Internet of things into a single information and material complex through a system of sensors, special-purpose devices and robotics. Agriculture digitalization involves the use of special software and computer applications, global positioning systems, robotics, self-driving aerial vehicles, artificial intelligence, block chain technologies, machine learning, etc. in production activities [5, 10].

The research made it possible to classify existing and promising digital technologies in agriculture by complexity level (Table 1).

The first group of technologies in the Table 1 includes the mobile applications and software for the producers to obtain information when making management decisions. The second group includes precision crop and livestock farming technologies using self-driving aerial vehicles, global positioning systems, the Internet of things and software based on the neural networks and machine learning. These technologies make it possible to increase all production processes efficiency, automate data accumulation and digitization and implement recommendation systems in agricultural enterprises management. The third group comprises self-driving agricultural machines, robotics, advanced achievements in the field of artificial intelligence. All mentioned things have not been widely used in agriculture, but have great potential for economic, technical and environmental efficiency [5].

Table 1. Classification of digital technologies in agriculture by complexity level

Technology complexity	Contents
Low	Information support with the use of industry mobile applications and software
Average	Precision crop and livestock farming technologies, self-driving aerial vehicles, geo-positioning and navigation systems, Internet of things, software with machine learning elements
High	Robotics, self-driving agricultural machines, artificial intelligence, digital platforms

3 Factors Influencing the Effectiveness of Russian Agriculture Digital Transformation

Wide dissemination of digital technologies in Russian agriculture, requires the activation of organizational, economic, technical, technological and social factors, which can be divided into basic and special ones (Fig. 1).

The basic factors ensure the beginning and further development of the program of Russian large-scale agriculture digitalization. The factors include the following issues:

- providing the agricultural producers with the access to high-speed Internet and mobile communication networks;
- human capital development in the rural areas;
- state support of the digital transformations processes in the agro-industrial complex;
- formation of a system of workers' professional retraining for the digital solutions and platforms operation;
- strengthening the agricultural producers' financial and economic condition.

The special factors include the following issues:

- leaders' and specialists' awareness of digital technologies economic benefits;
- innovative entrepreneurship development in the field of digitalization with import substitution of robotics and advanced achievements in the field of artificial intelligence;

– scientific foundations development for agricultural enterprises management in the frameworks of digitalization;
– creation of a reliable system of information and digital security and data protection.

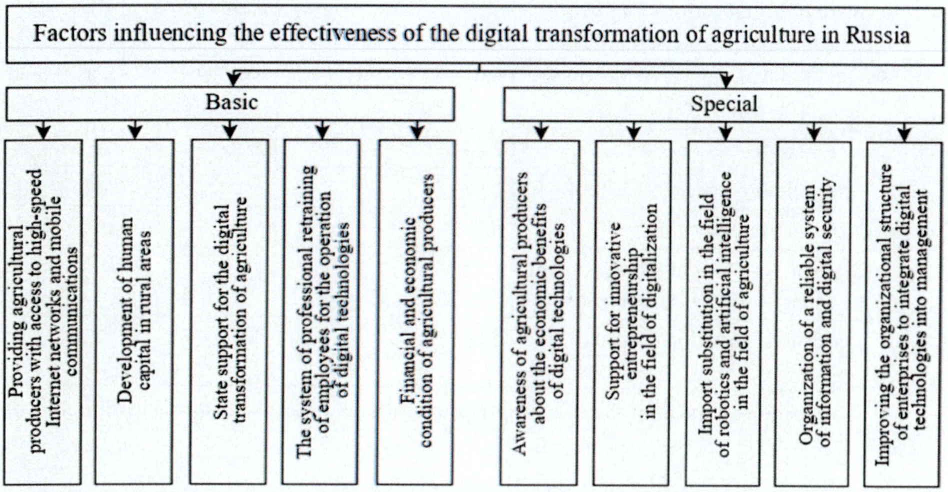

Fig. 1. Factors influencing the effectiveness of Russian agriculture digital transformation

4 Actual Status of Management System Digitalization for Technological Processes in Russian Agriculture

The analysis showed that currently the infrastructural conditions for large-scale agriculture digitalization have not been created in Russia. The main indicators of digital infrastructure development in the rural areas are presented in the Table 2.

Thus, the level of access to broadband high-speed Internet in the rural areas, where the agricultural organizations are mainly located, is significantly lower than in the cities. This fact creates one of the main problems for agricultural production accelerated digitalization. The rural population also has a relatively low level of digital literacy compared to the urban residents: only 5.8% of the rural population possess digital technologies above the entry level, while this indicator value is 14.3% in the cities [1].

At the same time, four Russian agricultural organizations out of five use the Internet only for sending e-mail. It is notable that after 2020, the agricultural organizations began using the Internet more actively to implement their employees' professional retraining. Currently about 25% of Russian agricultural organizations have their own websites. The most popular computer programs in the agricultural organizations are the systems of accounting and electronic document management used by more than 60% of agricultural enterprises.

Table 2. Key indicators of digital infrastructure development in the rural areas, 2019

Indicator	Agriculture in general	Crop farming	Livestock farming
General index of agriculture digitalization *	23	21	24
Share of organizations (%) using fixed broadband internet access	68.9	62.4	70.0
including data transmission speed above 100 Mbit/s	5.0	4.0	5.1
own website	24.1	23.6	24.3
electronic procurement system	14.3	н/д	н/д
electronic sales system	8.3	н/д	н/д
Share of agricultural workers using the Internet in their activities, %	10.5	11.5	9.5
Share of information security costs, %	2.8	3.2	2.0

* The arithmetic mean of the share of agricultural organizations using basic digital technologies; the source: [1]

According to the Higher School of Economics (HSE), the calculated index of agriculture digitalization of the Russian Federation was 23 percentage points in 2019. At the same time, its value in livestock farming was higher than in crop farming.

Krasnodar territory belongs to the regions with a high level of digitalization (its index was 29 p.p. in 2019). However, the share of workers regularly using digital technologies in the regional agricultural organizations is only 2.6%. It is the lowest value for all economic sectors [1].

According to the HSE, the costs of mobile communications and Internet services are the largest share in the costs structure of digital technologies usage in agriculture (59.9%). The cost of purchasing computer equipment and software is under 10%. At the same time, the share of agricultural organizations in total Russian investments in digital technologies creation and development is only 0.6%.

The costs of Russian agricultural organizations of software purchase and implementation amounted to about 1.0 billion rubles in 2019 (more than half accounted for the foreign developments). The software costs in livestock farming was 1.8 times higher than in crop farming.

According to the HSE, the agricultural enterprises are most interested in new production technologies (16%) and robotics (11%). Only 5% of organizations are interested in artificial intelligence and neural networks introduction. Other digital technologies are not of great interest to agriculture, and according to the HSE forecasts, the demand for them may decrease in the next 10 years (Fig. 2).

Currently, more than 74% of agricultural organizations have access to the broadband Internet. About 30% of agricultural organizations use the cloud services and only 5.5% – ERP-systems and RFID technologies. The electronic sales are established in 8.3% of agricultural enterprises. However, it should be noted that the digital technologies are used more actively by the livestock organizations.

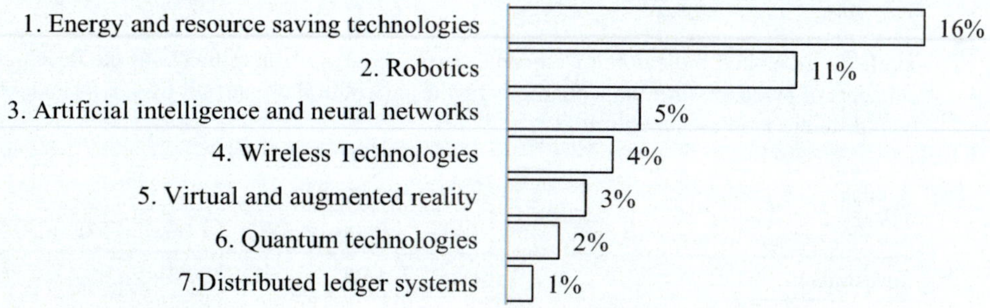

Fig. 2. Agricultural enterprises' demand for digital technologies, %

According to the Center for forecasting and monitoring in the field of precision agriculture of Kuban State Agrarian University, 189 farms of Krasnodar territory covering 963 thousand hectares used various elements of precision farming technology. It is about 20% of the total area of the regional agricultural land. However, the region mainly uses the simplest elements of precision agriculture technology, which do not require large investments from the commodity producers in purchasing of expensive complex devices and robotics. Thus, 156 regional agricultural enterprises use the technology of parallel driving of machine-tractor units, 94 – satellite monitoring of vehicle traffic, and 83 – mapping of fields according to productivity of cultivated crops. At the same time, it should be noted that the processes of mastering the elements of precision farming technology are implemented with varying intensity depending on different parts of the region and farms categories [9].

The investigations [9] have shown that among other things, Russian agriculture digitalization is restrained by the agricultural producers' weak awareness and distrust to the modern digital technologies. Thus, it is important to train farm managers and specialists on the issues of economic justification of the advantages of using precision agriculture technologies.

5 Expected Effectiveness and Riskiness of Investments in the Development of Precision Agriculture Technologies with the use of Digitalization Elements

Introduction of various elements of precision farming technology in the crop farming production makes it possible to improve the quality of field-mechanized operations, including sowing, soil and crops treatment, harvesting. This, in its turn, ensures the increase in the volume of manufactured products due to the growth of average crop capacity and reducing its cost by saving fertilizers and plant protection products.

The investigations have shown that the use of elements of precision farming technology can increase crop capacity by 15–20%, reduce fuel costs by 25–30% and mineral fertilizers costs by 15–20%.

The estimated indicators of efficiency and riskiness of investments in the development of elements of precision farming technology by the agricultural organizations of Krasnodar territory with different land use sizes are presented in the Table 3.

Table 3. Estimated indicators of efficiency and riskiness of investments in the development of elements of precision farming technology by the agricultural organizations of Krasnodar territory with different land use sizes, mln. RUB

Indicator	Land use size, ha					
	100	300	500	1 000	3 000	5 000
Investments	14.4	15.4	17.4	30.3	49.4	64.1
Expected annual economic effect	1.6	4.7	7.9	15.8	47.4	79.0
Net discounted income	−8.9	0.9	9.7	23.9	113.3	207.0
Internal rate of return, %	−17	16	36	44	92	121
Payback period of investments, years	–	4.6	2.8	2.4	1.2	0.9
Project loss possibility, %	100	34.9	0.1	0	0	0

The investments in development of the elements of precision farming technology by the agricultural organizations with the land use sizes up to 300 hectares are effective and repaid in less than 5 years. However, in this case, the investment loss possibility is quite high (35%).

The research has determined that the risk of project loss possibility is reduced to a minimum for the commodity producers with land use sizes of more than 500 hectares, and the capital investments are repaid in less than 3 years. The further growth of the land use size significantly increases the net discounted income and the internal rate of return for the considered projects, and decreases the discounted payback period of investments.

The analysis of the level of investment riskiness in the implementation of considered technological innovation has been carried out by the method of simulation modeling for the regional agricultural organizations with the land use sizes of 300 and 500 hectares. It allowed constructing graphs of distribution frequency of the net discounted income and determining the investment break-even possibility in these innovative projects implementation (Fig. 3).

The results obtained have shown that the investments break-even probability in the considered project implementation for a model agricultural organization with the land use size of 300 hectares is about 65%. Such projects are practically risk-free in the agricultural organizations with the land use size of more than 500 hectares.

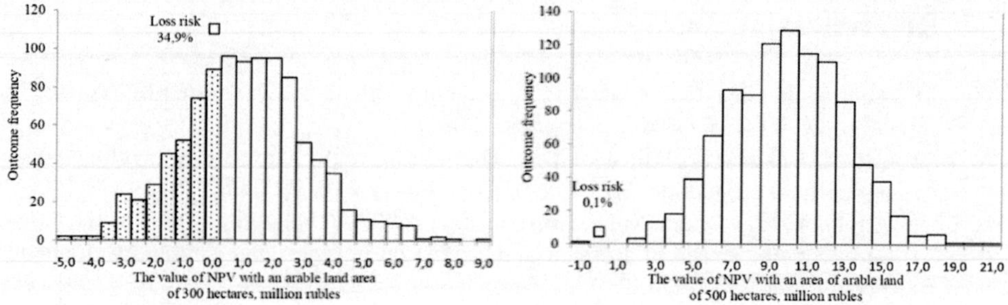

Fig. 3. Distribution frequency of the net discounted income of an innovative investment project for the development of the elements of precision farming technology with different arable land areas by the agricultural producers

6 Conclusion

Successful problem solving of import substitution of agricultural raw materials and food in the national market, ensuring Russian food security and increasing its export capacity will largely depend on the national agriculture ability to begin using the innovative agricultural technologies based on extensive automation, robotization and informational support of production processes with the elements of digitalization and artificial intelligence.

The priority direction for such innovative transformations implementation is widespread introduction of precision crop and livestock farming technologies into Russian agriculture. Currently, Russian agriculture digitalization is mostly restrained by insufficient access to the broadband Internet by the national agricultural producers.

Only 189 agricultural organizations out of more than 600 with land use sizes amounting to about 20% of the total area of farmland in Krasnodar territory, one of the leading agricultural Russian regions, currently implement the individual elements of precision farming technology. The analysis of production activity of these advanced economic entities shows that the application of precision farming technologies allows increasing the average cultivated crops capacity through targeted and differentiated fertilization and the use of chemical protection agents, as well as reducing current production costs, which increases the production profitability and competitiveness.

The performed calculations have shown that efficiency and riskiness of investment in the development of precision farming technology with the elements of production processes digitalization in Krasnodar territory is mostly determined by the land use size of an agricultural organization. The investments in a project sized of less than 300 hectares are highly risky and economically unjustified. At the same time, if the land use area of an agricultural organizations exceeds 500 hectares, the project loss possibility is reduced to a minimum, and the payback period of investments does not exceed 3 years.

Acknowledgments. The research has been implemented with the financial support of Kuban Science Foundation as a part of the research project no. 20.1/73.

References

1. Abdrakhmanova, G.I., et al.: Digital economy indicators: 2021: statistics collection. Russia, Moscow: Vysshaia shkola economiki (2021)
2. Alt, V., Isakova, S., Balushkina, E.: Digitalization: problems of its development in modern agricultural production. E3S Web of Conferences **210**, 10001 (2020)
3. Altukhov, A.I., Dudin, M.N., Anishchenko, A.N.: Global digitalization as the organizational and economic basis of Russian agro-industrial complex innovation development. Market Econ. Probl. **2**, 17–27 (2019)
4. Bershitskii, Iu.I., Saifetdinov, A.R., Puzeichuk, P.V.: Organization of innovation activity in the agro-industrial complex. Russia, Krasnodar: KubGAU (2019)
5. Bolfe, E.L., et al.: Precision and digital agriculture: Adoption of technologies and perception of Brazilian farmers. Agriculture **10**, 653 (2020)
6. Faskhutdinova, M.S., Amirova, E.F., Safiullin, I.N., Ibragimov, L.G.: Human resources in the context of digitalization of agriculture. BIO Web of Conferences **27**, 00020 (2020)
7. Saifetdinov, A.R., Saifetdinova, P.V.: Basic economic principles of the digital transformation in Russian agriculture. IOP Conference Series: Earth and Environmental Science **745**, 12009 (2021)
8. Shepherd, M., Turner, J.A., Small, B., Wheeler, D.: Priorities for science to overcome hurdles thwarting the full promise of the 'Digital Agriculture' Revolution. J. Sci. Food Agric. **100**, 5083–5092 (2018)
9. Trufliak, E.V.: The use of precision agriculture elements in Russia. Russia, Krasnodar: KubGAU (2018)
10. Ulezko, A., Reimer, V., Ulezko, O.: Theoretical and methodological aspects of digitalization in agriculture. IOP Conf. Series: Earth and Environmental Science **274**, 012062 (2019)

Evaluation of the Efficiency of State Support for Sustainable Development of Agriculture in the North Caucasian Federal District

Igor Kaschaev(✉) and Elena Kriulina

North Caucasian Federal Scientific Agrarian Center, Mikhailovsk, Russia
kiv_76@mail.ru

Abstract. The need to adapt to the increasingly complex economic reality makes state regulation of the sustainable development of agriculture extremely relevant. The improvement of agricultural production efficiency and support of a decent standard of living for the rural population requires the combination of various methods, techniques, forms and tools. World experience shows that in market conditions, stable agriculture is not able to function successfully without serious permanent state support. The agrarian economy of the North Caucasus Federal District is characterized by growth without development, which, in the context of an inefficient economic structure, significantly limits its effectiveness. Agriculture is a backbone sector of the region's economy and requires an increase in both funding and efficiency of the use of allocated funds, as well as an increase in the scale of application of innovations in the industry. However, due to the growing deficit of the Russian budget, the state can currently take over only part of the costs by financing the most important areas of development of the agro-industrial complex. Therefore, the goal of the economic mechanism of the development of agriculture is to achieve compliance between the tasks set for the industry and the system of interconnected economic instruments that ensure their implementation, as well as the available and allocated resources. The article presents the dynamics and structure of state support for agriculture in the Russian Federation and the North Caucasus Federal District from 2015 to 2020. An analysis and assessment of the state, trends, and effectiveness of state support for the intensification of the agricultural sector were carried out. A conceptual model of state regulation of the sustainable development of agriculture in the North Caucasus Federal District is proposed.

Keywords: Sustainable development · Economy · Agriculture · Efficiency of state expenditures · Subsidies

1 Introduction

The state policy of financial support for agriculture in 2008–2020 was changed and updated almost every year. Directions, forms, mechanisms, and amount of state support were corrected, usually associated with a decrease in its volume. The most significant

I. Samoylenko (Ed.): AMBT 2022, LNNS 582, pp. 73–82, 2023.
https://doi.org/10.1007/978-3-031-20803-4_8

changes occurred in 2013 (introduction of "unrelated income support" of crop production, economically significant regional projects); in 2017 (federal transfers began to operate in the form of a "single subsidy"); and also, in 2020 (introduction of "stimulating" and "compensating" subsidies). Structural changes in the state support system have led to a transition from sectoral subsidies (direct support measures) to aggregated support measures (introduction of a "unified subsidy") and to a shift in emphasis to project and process management of the industry, including projects and targeted programs that determine the vector of state support for the next five years. Today, the mechanisms for providing state support to the agro-industrial complex (AIC) have become as transparent and predictable as possible, creating a stable basis for long-term planning and supporting the activities of agricultural producers in their constant intensive sustainable development.

The study is based on a systematic approach using general logical methods and techniques, as well as monographic, economic-statistical, analytical, and other methods. The information base of the study includes government programs, analytical materials, as well as regulatory, reporting, and information documentation of the ministries of agriculture and economic development of the Russian Federation and the subjects of the North Caucasus Federal District. The empirical basis of the study is the materials of periodic and operational statistical reporting, generated by the Office of the Federal State Statistics Service for the North Caucasus Federal District.

2 Analysis of State Regulation of the Stable Development of the Agricultural Sector in Russia and the North Caucasus Federal District

In the economic literature, there are interpretations of the terms "state regulation", "state aid" and "state support". The broadest concept is state regulation, including state support at the macro-, meso-, and micro- levels, protectionist, price, tax, credit, investment policy, measures for the financial recovery of organizations, etc. State support for agriculture has a significant impact on economic growth in the agricultural sector of the North Caucasus Federal District, so the choice of its most effective directions is of paramount importance. State support is a set of funds and benefits received free of charge or on preferential terms by agricultural producers and rural areas from the budgets of different levels for effective development. Its objects can be individual industries, business entities, certain processes, as well as specific territories [7, 11, 14].

The dynamics and structure of state support for the industry in the Russian Federation (RF) and separately for the North Caucasian Federal District for 2015–2020 are presented in the Table 1, compiled by the authors based on data from the Federal State Statistics Service (Rosstat) and the Office of the Federal State Statistics Service for the North Caucasus federal district (North Caucasus State Statistics) [2, 3, 8].

The average annual share of federal budget funds in the total amount of financing for the industry in the Russian Federation was 75%, the corresponding figure for the North Caucasus Federal District was almost 88%, from 80% in the Stavropol Territory to 93% in the Chechen Republic.

Table 1. Dynamics and structure of state support for agriculture

Years	Russian federation			North caucasian federal district		
	Consolidated budget, billion rubles	Share of budgets, %		Consolidated budget, billion rubles	Share of budgets, %	
		Federal	Regional		Federal	Regional
2015	220.1	76.3	26.7	15.1	85.8	14.2
2016	211.7	73.4	26.6	14.7	86.6	13.4
2017	203.6	70.7	29.3	15.2	86.7	13.3
2018	206.5	73.3	26.7	15.7	88.6	11.4
2019	203.9	72.5	27.5	15.3	87.9	12.1
2020	243.3	83.5	16.5	17.6	93.8	6.2
Average	214.9	75.0	25.0	15.6	87.9	12.1

The amount of budgetary (federal and regional) expenditures aimed at supporting agriculture in the North Caucasus Federal District in 2020 is 17.6 billion rubles, or 7.2% of its total volume in the Russian Federation. The large difference in the share of regional budgets in Russia as a whole and in the North Caucasus Federal District (16.5% and 6.2%, respectively) draws attention. In the structure of state support for the subjects of the district under agreements with the Ministry of Agriculture of Russia, the largest share is occupied by a compensatory subsidy (24.3%), an incentive subsidy (22.4%) and subsidies for the integrated development of rural areas (20.6%). The volume of subsidies transferred from the regional budgets in the North Caucasus Federal District amounted to 1,084.4 million rubles. The structure of state support from regional budgets (in addition to agreements with the Ministry of Agriculture of Russia) is mainly represented by subsidies for the development of infrastructure in rural areas (26.0%), investments in fixed assets (24.2%), compensation for current costs in crop and livestock production (14.5%), subsidies to reimburse part of the interest rate on investment loans (loans) in the agro-industrial complex (6.3%) [1, 3].

Support for the intensification of the agricultural sector of the district will remain practically at the same level in the next five years (2021–2025). Thus, the amount of support for 2022 is determined at the level of 15.6 billion rubles, which is insufficient both from the standpoint of ensuring conditions for expanded reproduction and the contribution of the agricultural sector to the all-Russian production of gross agricultural output. On average, it amounted to 8.4%, while the share of funds to support the industry turned out to be only 7.2% [1].

3 The Effectiveness of State Regulation of Priority Subsectors for Sustainable Development of the Agro-Industrial Complex

Modern directions for the implementation of state regulation of the stability of the agro-industrial complex until 2025 include four targets:

1. Ensuring food independence following Food Security Doctrine of the Russian Federation.
2. Achievement of indicative (actual) values of the produced added value in agriculture.
3. Achieving the volume of exports of agricultural products in comparable prices up to 35.4 billion dollars.
4. The index of the physical volume of investments in fixed capital of agriculture (excluding small businesses) in 2025 should be 118.2% compared to 2017.

In 2021, the number of areas of support was reduced to two: the development of the agro-industrial complex and the ensuring of conditions for the functioning of the agro-industrial complex. They include projects and targeted programs that determine the vector of state support for the next five years [1]. These include the following:

1. Federal projects.
1.1. Export of agricultural products.
1.2. Creation of a support system for farmers and development of rural cooperation.
2. Departmental projects.
2.1. Development of sectors of the agro-industrial complex that provide accelerated import substitution of the main types of agricultural products, raw materials, and food.
2.2. Stimulation of investment activity in the agro-industrial complex.
2.3. Technical modernization of the agro-industrial complex.
2.4. Digital agriculture.

The structure of priority areas for the development of the agro-industrial complex in the constituent entities of the Russian Federation includes five sub-sectors of crop production, four sub-sectors of livestock farming, and separately - small farms. The analysis showed that all regions consider the development of small farms as the priority

Table 2. List and distribution of priority sub-sectors of agriculture in the constituent entities of the Russian Federation in 2021–2023

No.	Priority sub-sectors	Entities of the RF, units
1	Production of cereals and leguminous crops	58
2	Production of oilseeds (excluding rapeseed and soybeans)	33
3	Open field vegetable production	30
4	Development of viticulture	11
5	Manufacture of products of fruit and vegetable plantations, including the care of perennial plantations	53
6	Production of fiber flax and (or) industrial hemp	11
7	Milk production	63
8	Production of specialized beef cattle breeding	35
9	Development of small farms	82

direction. (Table 2) [5]. This form of entrepreneurial activity needs state support, but in the absence of a federal project and the financial resources provided for in it, the development of rural cooperation is problematic.

Highly demanded sectors for the period 2021–2023 are the production of grain and leguminous crops and the production of milk: these sub-sectors are identified as a priority in the development programs by 58 and 63 constituent entities of the Russian Federation, respectively.

The subjects of the district identified production of fruit and berry plantations, sheep and goat breeding, development of small farms as priority sub-sectors. (Table 3).

Studies show that for economic growth in agriculture direction in which financial resources are spent (for general services or for supporting individual producers) is more important than the overall level of support. Reallocation of 10% of the budget from subsidies to support general services leads to a 5% increase in value-added in agriculture [6]. These include programs and activities aimed at creating general conditions for managing and increasing the potential of the industry: support for science, education, inspection and control services, development of agricultural infrastructure and rural development, marketing development, etc.

Several studies have revealed a negative correlation between budget expenditures to support agricultural producers and economic growth in agriculture [11, 14]. The opinions that support producers directly through subsidies and benefits lead to the crowding out of the private investment, deprives incentives to increase competitiveness, creates a burden on taxpayers, and distorts market signals are well-reasoned.

At the same time, measures focused on common services smooth out market distortions caused by subsidies [7, 12].

A comparative assessment of the effectiveness of the use of state financial support by the regions of the North Caucasian Federal District is presented in Table 3 compiled by the authors based on data from Federal State Statistics Service and North Caucasus Statistics Service. [2, 3, 9]. The effectiveness of budget support for agriculture in the North Caucasus Federal District is determined by the effectiveness of a set of measures for industry-wide support for agricultural producers of all organizational and legal forms aimed at improving economic, financial and social sustainability, stimulation of the production of certain types of agricultural products, ensuring productive employment of the rural population. The calculation of the payback of the allocated state support funds by agricultural products allows us to assess the rationality of their use in the subjects of the North Caucasus Federal District [13].

Table 3. Priority areas for sustainable development of the agro-industrial complex of the subjects of the North Caucasus Federal District

Priority subsectors	Dagestan	Ingushetia	Kabardino-Balkaria	Karachay-Cherkess	North Ossetia–Alania	Chechen Republic	Stavropol Krai
Crop production							
Production of cereals and leguminous crops	☑	☑	☑	☑	☑		☑
Production of oilseeds (excluding rapeseed and soybeans)	☑		☑	☑	☑		☑
Open field vegetable production	☑		☑		☑		
Development of viticulture	☑	☑		☑	☑	☑	☑
Manufacture of products of fruit and berry plantations, including the care of perennial plantations	☑	☑	☑	☑	☑	☑	☑
Livestock							
Milk production	☑	☑	☑	☑	☑	☑	
Development of specialized beef cattle breeding	☑		☑	☑	☑	☑	
Development of sheep and goat breeding	☑	☑	☑	☑	☑	☑	☑
Development of small farms	☑	☑	☑	☑	☑	☑	☑

It follows from the data in the table that in 2020, on average in the district, one ruble of state support turned into 28.7 rubles of gross agricultural output. In three of the seven subjects of the North Caucasus Federal District, the level of payback of state support did not reach the average level (Table 4).

It is especially low in the Republic of Ingushetia (10.2 rubles). The payback of state support in the Kabardino-Balkarian Republic and the Chechen Republic was a little bit higher, but lower than the average level for the district. The fact that in Stavropol Krai for 901 rubles per 1 hectare of the state support, the total volume of agricultural production was 1,2 times more than the average volume in the district is noteworthy. Several conclusions can be drawn from this fact. Firstly, the effectiveness of the use of state support funds by the regions of the district varies significantly, in the Stavropol Territory it is the highest. Secondly, due to the acute long-term insufficiency of funds for the sustainable development of agriculture, the financial resources allocated today in the form of state support are clearly not enough. Thirdly, in the three subjects of the district with a low return on state support funds and the absence of a direct connection between state support measures and its results, substantive analysis of the directions of their use is

Table 4. The volume, level and effectiveness of the use of funds allocated to the regions in the form of state support for agricultural production in 2020.

Regions	The total amount of state support, million rubles	Area of agricultural land, thousand hectares	State support per 1 hectare, thousand rubles	Volume of agricultural production, billion rubles	Payback of state support for industry products, rubles
North Caucasian Federal District	17590.1	13534.6	1299	504.3	28.7
Republic of Dagestan	4215.6	4344.3	970	141.5	33.6
Republic of Ingushetia	1116.6	150.9	7399	11.4	10.2
Kabardino-Balkarian Republic	2751.6	711.0	3870	61.4	22.3
Karachay-Cherkess Republic	1031.4	815.9	1264	31.4	30.4
Republic of North Ossetia–Alania	1184.6	418.6	2829	39.2	33.1
Chechen Republic	1791.0	992.3	1804	33.3	18.6
Stavropol Krai	5499.3	6101.6	901	186.3	33.9

needed in order to approach the distribution of allocated funds more rationally. Fourthly, it is necessary to consider and implement as much as possible the opinion of the rural population and the agrarian business, which was incorporated in the development of the Social and Economic Development Strategies by the subjects of the region on the directions for investing the funds allocated by the state.

4 Model of State Regulation of Sustainable Development of Agriculture in the North Caucasus Federal District

An integrated approach to state regulation of agricultural intensification involves structuring actions, supporting processes at all stages, synchronizing funds, improving program-targeted methods, evaluating efficiency, and updating. The authors propose a model of this systemic interaction of tasks, instruments, and resources, which is shown in Fig. 1.

To eliminate negative trends in the North Caucasus Federal District, it is necessary to actively develop new technological industries in the agricultural sector, create high-performance jobs with decent wages, upgrade the logistics infrastructure to expand product markets, and form value chains. All these are the elements of sustainable development of agriculture. Therefore, an important condition for increasing the economic efficiency of agricultural production and achieving the stability of agricultural systems is the intensification of investment activity in the agriculture of the region. This will create the necessary conditions for increasing the level of technical equipment, the economic efficiency of agricultural production by improving the composition and structure of fixed assets, and, ultimately, ensuring sustainable development.

To improve the efficiency of budget financing, it is advisable to use a selective electoral mechanism of state support. For each subject of the North Caucasus Federal District, it is important to identify the most significant priorities that require investment

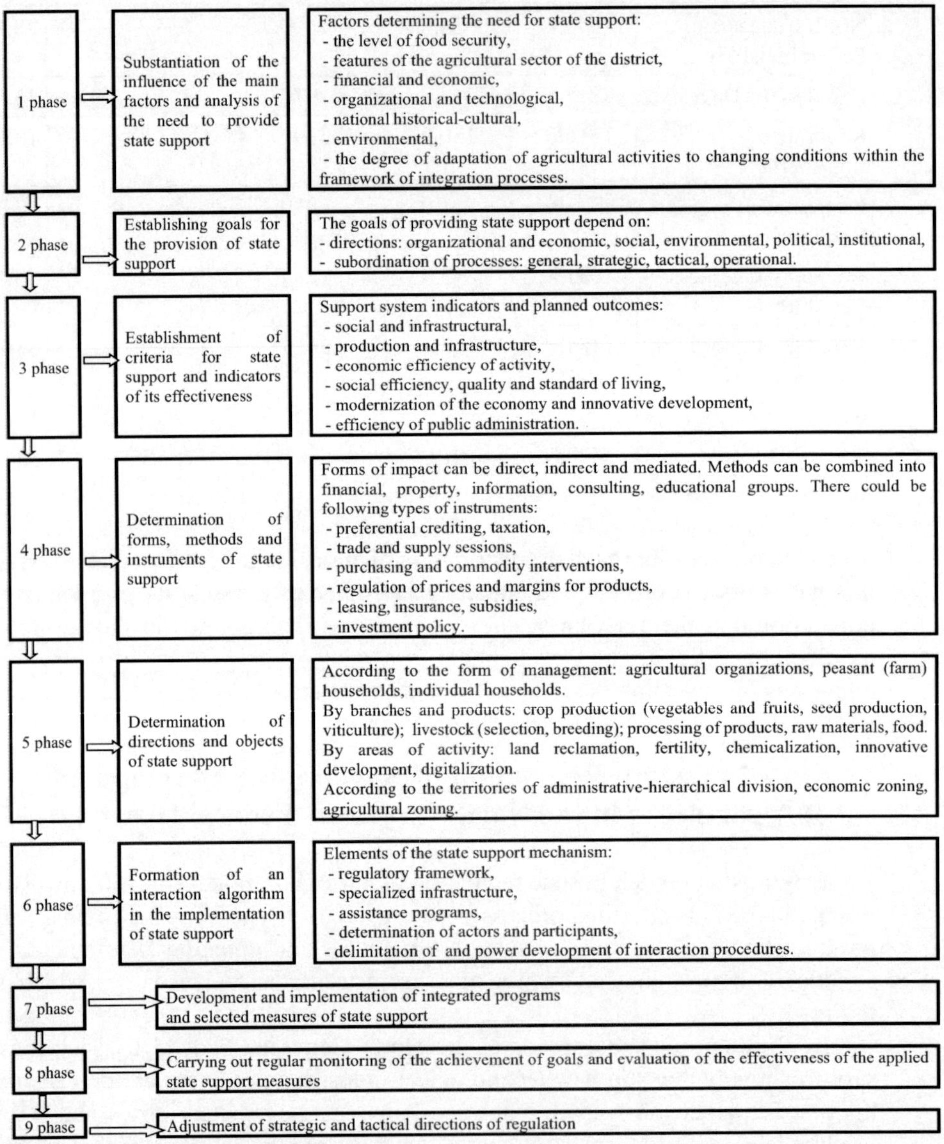

Fig. 1. Conceptual model of state regulation of sustainable development of agriculture in the North Caucasus Federal District

support (sustainable development of rural areas and improving the living standards of the rural population, increasing the competitiveness of agricultural production and the reproduction of agricultural resources used, etc.). At the same time, it is necessary to consider the specifics of the available resources and risks in each specific local area [10]. For investment-attractive business entities, state support should aimed at developing the potential of large organizations, stimulating the introduction of innovations. For potentially investment-unattractive entities, state support should be focused on the effective use of the potential of small and medium-sized businesses, reducing investment risks. For investment-depressed territories, a model of focal development is needed, which implies the concentration of production on available resources around infrastructure facilities.

In addition, the directions of development of the branches of the agrarian economy and the amount of funding should consider the interests of different groups of producers in the conditions of the multi-structural economy of the North Caucasus Federal District [4]. Currently, state support is focused mainly on large-scale agricultural production, while the conditions for agricultural producers are constantly changing.

5 Conclusion

It should be taken into account that the diversification of budget support, being an effective tool for expanding activities and increasing production volumes, can also lead to a slowdown in the reproduction processes in agriculture. Therefore, budget support for the industry must be carried out in conjunction with other elements of the state support mechanism. At the same time, it is necessary to consider the high degree of differentiation of the subjects of the North Caucasus Federal District in terms of the main socio-economic parameters, the share of the agricultural sector in the gross regional product, competitive advantages in the production of certain agricultural products, etc. These factors, as well as maximum transparency, openness and predictability of budget expenditures, must be taken into account when improving and increasing the efficiency of state support, one of the most important elements of sustainable development of agriculture in the North Caucasus Federal District.

References

1. Government of the Russian Federation: State program of the Russian Federation "Development of the North Caucasian Federal District". Approved by Decree of the Government of the Russian Federation of April 15, 2014 (No. 309) (2014)
2. Unified interdepartmental information and statistical system: Electronic resource. EMISS. State statistics. URL: https://www.fedstat.ru/organizations/
3. Ministry of Agriculture of the Russian Federation: Information guide on the measures and directions of state support for the agro-industrial complex of the Russian Federation
4. Kashchaev, I.V., Kriulina, E.N., Kriulina, E.N.: Conceptual mechanisms for increasing the sustainability of agricultural activities of the subjects of the North Caucasian Federal District. APK: Economics, management, No. 1 (Agrarian policy: problems and solutions), pp. 3–8 (2022)

5. Ministry of Agriculture of the Russian Federation: Order of the Ministry of Agriculture of the Russian Federation No. 674 dated November 11, 2020 “On determining priority areas for the development of the agro-industrial complex for 2021, 2022 and 2023” [Electronic resource]. Code (2020). URL: https://docs.cntd.ru/document/499091772
6. Perova, A.E.: Evaluation of the effectiveness of state support for domestic agricultural producers. International research journal, No **04–1**(58), 152–157 (2017)
7. Rodionova, O.A.: Intersectoral distribution of budgetary support in the implementation of the project-process approach. APK: Economics, management, No. 5 (Economic mechanism of management), pp. 38–43 (2021)
8. Federal State Statistics Service: Socio-economic situation of the federal districts [Electronic resource]. URL: https://gks.ru/bgd/regl/b20_20/Main.htm
9. Office of the Federal State Statistics Service for the North Caucasian Federal District (2022). URL: https://stavstat.gks.ru/
10. Shelkovnikov, S.A., Petukhov, M.S.: Evaluation of the effectiveness of state support for the technical modernization of agriculture. Economics and business: theory and practice **1**, 122–125 (2015)
11. Shik, O.V., Serov, E.V., Yanbykh, R.G.: Study of the system of budget support for the agricultural sector in Russia. Issues of state and municipal management **2**, 145–167 (2020)
12. Aliev, F.M., Imashova, D.G., Sheykhova, P.M., Kamalova, P.M.: Regional Peculiarities of State Support for Agriculture. In: Popkova, E.G. (ed.) Russia and the European Union. CE, pp. 69–74. Springer, Cham (2017). https://doi.org/10.1007/978-3-319-55257-6_10
13. Glotova, I.I., Tomilina, E.P., Klishina, Y.E., Uglitskikh, O.N., Agarkova, L.V.: The current state of the state support for agriculture in the region. Res. J. Pharmaceutical Biological and Chemical Sci. **9**(6), 1926 (2018). Prodatur, 00000, India. https://www.elibrary.ru/item.asp?id=37103052
14. Neganova, V.P., Dudnik, A.V.: Improving the state support of agriculture in a region. Ekonomika Regiona-Economy of Region **14**(2), 651–662 (2018). RUSSIAN ACAD SCIENCES, URAL BRANCH, INST ECONOMICS. https://doi.org/10.17059/2018-2-25

Youth Commitment to Traditional Ethical and Moral Values as a Factor of the Regional Community's Sustainability

Maxim Abramov[1], Tatiana Popova[2], Andrey Kurnosenko[3], Dmitry Skvortsov[4], and Valentina Ivashova[5](✉)

[1] Essentuki Institute of Management, Business and Law, Essentuki, Russia
6225353@gmail.ru
[2] Academy of Management of the Ministry of Internal Affairs of Russia, Moscow, Russia
[3] Main Directorate of Economic Security and Anti-Corruption, Moscow, Russia
[4] Nevinnomyssk State Humanitarian and Technical Institute, Nevinnomyssk, Russia
[5] Stavropol State Agrarian University, Stavropol, Russia

Abstract. The sustainable development of the regional socio-economic system is determined by a wide range of factors, including goal-setting, strong-willed efforts and the willingness of various groups of the regional community to support this or that activity. Spiritual and moral values are the normative basis and regulator of this conscious activity and behavior of people. The assessment of the level of commitment of young people to traditional spiritual and moral values is relevant for understanding the degree of stability of the regional community and is the purpose of our research. A theoretical review of the subject field associated with the preservation of spiritual and moral values of various regions of the world and population groups confirms the relevance of the discussion track in the research of the modern scientific community. The results of an empirical research of the level of youth commitment to traditional spiritual and moral values showed a high level of preservation and reproduction in the student environment of future specialists in the agrarian sector of Stavropol Region (Russia). The reproductive potential of traditional spiritual and moral values contributes sustainability to the development of the regional community.

Keywords: Regional community · Traditional ethical and moral values · Youth commitment · Regional community's sustainability

1 Introduction

Sustainable development of the regional socio-economic system is determined by a wide range of factors. Among them, the social component plays a key role, since the actualization of any processes depends on the goal-setting, will and readiness to support the proposed actions of certain groups of the regional community. We are talking about the leaders of the regional community (heads and specialists of the region, municipalities,

I. Samoylenko (Ed.): AMBT 2022, LNNS 582, pp. 83–90, 2023.
https://doi.org/10.1007/978-3-031-20803-4_9

enterprises and organizations), ordinary workers, and other social groups (youth, students, pensioners, housewives, etc.). Ethical and moral values are just the normative basis and regulator of this conscious activity and behavior of people. Separate attitudes of the individual are formed in the system of the most important social institutions of society (family and education) and lay the cultural, moral, social meanings that orient a person's actions towards the achievement of higher ideals. In the process of reproduction of traditional spiritual and moral values, two complementary aspects are important – preservation and development in accordance with the changing context of the historical period. Understanding the content side of the reproduction process of the ethical and moral values of the regional community, which ensure the sustainability and development of society, is possible through studying the level of commitment to traditional moral values of today's youth. Particular attention in organizing a sociological assessment should be given to student youth, which, on the one hand, is the most active part of the regional community and forms an innovative component of the existing ethical and moral traditions. On the other hand, in the near future they will be representatives of the professional community of specialists, heads of structural divisions and organizations who demonstrate the role model to their subordinates. And the sustainable development of production teams, as a component of the regional socio-economic system, also depends on the content of this role model. The issue is of particular relevance when it comes to an agrarian region with a traditional rural way of life, features of labor objects and the nature of production processes in the agricultural industry. Thus, the assessment of the level of youth commitment to traditional ethical and moral values is relevant for understanding the degree of stability of the regional community and is the purpose of our research.

Various aspects of the subject field associated with the preservation of the ethical and moral values of various regions of the world and population groups are an active discussion track in the research of the modern scientific community. The practices of maintaining professional traditions of social responsibility, which are implemented in the management of the financial resources of a family business, are shown in the publication of the authors N.B. Sah, A. Banerjee, J. Malm, A. Rahman [1]. Based on the results of a study on the example of several family firms, it was found that management approaches have a certain similarity: avoidance of risky financial policies, commitment to business sustainability and reputation, conservative short-term investment policy, timely payment to suppliers. And these approaches in organizing a family business are becoming relevant professional practices for the younger generation and ensure the preservation of not only the traditions of production management and financial processes, but also traditions related to the ethical and moral sphere. These are life wisdom, social responsibility, the value of professional knowledge, the priority of product quality and production sustainability over financial gain, continuity, diligence, honesty. In general, the preservation of the traditions of family business ensures sustainable socio-economic development of the territories where such firms operate. The results of the research confirm the relevance of studying the commitment of young people to traditional ethical and moral values. The preservation of the ethical and moral health of the younger generation is ensured by the functions of the social institution of the family. And the protection of children from modern information challenges depends on how well these functions are performed. One of them is the oversight function. The authors of the study N. Atman Uslu, H. Yildiz Durak, based on a survey of the parent community and subsequent factor analysis of the data, prove that the security of the ethical and moral health of the child depends not only on the sustainability of the reproduction of ethical and moral values in the family, but also on the level of family supervision over the content of information communications, timely protection against possible situations of cyber bullying [2]. In addition, the study found that there is a dependence of the preservation of family supervision functions on the age of the parents (the lower the age, the less the function manifests itself). Therefore, the significance of continuity in the reproduction of ethical and moral values and family traditions is acquiring new contours in connection with the challenges of our time. Thus, adherence to family traditions is associated with a reduction in social contradictions and conflict situations and gives social sustainability to the local community. These are important findings for our research, which confirm its relevance and significance.

The period of the global COVID 19 pandemic, like any crisis situation, clearly shows hidden processes and different response strategies, showing the ethical and moral basis of the traditional way of life of different social groups and regions. The authors of the study (D. Joseph, N. Belford, R. Lahiri-Roy) use a number of national cases to show how remote assistance from children to elderly parents helps to survive the difficulties of a pandemic [3]. Respect for family values through financial assistance to parents brings additional socio-economic sustainability to the region where parents are located, care for which is not shifted to the municipality. We see a positive relationship between the preservation of family values of responsibility, care for the older generation and a certain contribution to the socio-economic sustainability of the municipality and the local community.

The cultural values of parents have a significant impact on the formation of sense of community in a group of children. This relationship is traced in their study by the authors of the article A. Gampe, J. Blaumeiser, M.M. Daum [4]. In the course of studying the level of loyalty in the children's team through the mechanism of parental influence on the processes of identification and attitudes towards the social group, a positive effect on the formation and maintenance of groups, sense of community within the social group was established. In turn, this testifies to the contribution of the older generation to the normative regulation of loyalty in society, the preservation of the function of the family as an ecological cultural niche for the transfer of social experience and the value of good neighborly, conflict-free relations.

Family values, including those related to the issues of procreation, are reflected in the publication of the authors J. Kim, S. Joo, K. Lee, H. Jun [5]. The preservation of family values is a significant part of the traditional ethical and moral values and ensures the demographic stability of the local and regional community. In general, such a position makes a positive contribution to understanding the social stability of the regional community through the mechanism of preserving traditional spiritual and moral values [6–8].

Thus, the brief literature review on the issue of affecting the preservation of traditional ethical and moral values shows the relevance and significance of our research [9–13].

2 Materials and Methods

A survey of respondents from among the student youth of Stavropol State Agrarian University was conducted in December 2021 (Stavropol, Russia) using the handout questionnaire method. In total, 423 people took part in the study, who are students of 2–3 year and study full-time on the programs of Agronomy, Veterinary Medicine, Economics, Ecology, Engineering, Service and Tourism. The participation of students is organized with a quota representation of these programs of training. The ratio of student youth by gender is also maintained. In the toolkit for studying the level of youth commitment to traditional ethical and moral values, 3 information sections were identified that structure empirical indicators: Sect. 1 – the preservation of ethical and moral traditions in the family; Sect. 2 – the development of value orientations in the youth environment and their sustainability; Sect. 3 – internationality of ethical and moral according to the youth community. The survey database was processed in SPSS Statistics (version 21). Statistical procedures made it possible to analyze the results of the study in segmentation by independent variables: family characteristics, nationality, and permanent place of residence. Assessing the level of youth commitment to traditional ethical and moral values makes it possible to use information to improve educational work, activities of the socio-psychological service of the university and tutors of study groups. The article presents the most significant information obtained during the study, which shows the contribution of young people's commitment to traditional spiritual and moral values to the social stability of the regional community.

3 Results and Discussion

Distribution of answers to the questions of Sect. 1 – the preservation of ethical and moral traditions in the family. In the course of a survey of student youth, it was found that the majority of survey participants (mentioned by 64.8% of respondents) are well aware of the history of their family (where they lived and who their ancestors were, etc.). 35.2% of respondents know family history, but not very well. The options "I don't know" or "I can't say for sure" were not chosen by the survey participants. 81.6% of respondents indicated that the family has traditions that are passed down from generation to generation, which indicates a high level of reproduction of traditional family values. 18.4% of respondents believe that their families do not have such traditions. A good knowledge of family traditions was mentioned by the majority of survey participants – 80.4%; not very good knowledge of family traditions was mentioned by 19.6% of respondents. Opinions about the personal attitude of young people to the observance of family traditions were divided almost equally: 51.5% of respondents indicated that they adhere to all family traditions and 48.5% try to adhere to some traditions. 19.9% of respondents noted that recently the family has become more responsible for family traditions. 16.8% of respondents noted that recently the family has become less responsible for family traditions. 50.3% of respondents indicated the stability of traditions in the family – over time, they are observed as strictly as before. And 13.0% of the survey participants found it difficult to give a definite answer. The reproductive process of family traditions and ethical and moral values is undergoing changes, the question arises of their relevance in modern conditions. According to the majority of representatives of the student community, all and most of the family traditions are relevant in modern conditions (37.8% and 48.5% of respondents, respectively). Only a small part – 1.2% of the survey participants noted that family traditions have lost their relevance and 12.5% of the survey participants could not give a definite answer.

Distribution of answers to the questions of Sect. 2 – the development of value orientations in the youth environment and their sustainability. Preservation of family values is the basis of normative regulation in making vital decisions. Data on the impact of traditional values on family formation and family life are presented in Table 1.

Table 1. Presence of traditional values in family life, %

	Yes, definitely	No, not necessary	Difficult to say
1. Mandatory consent (blessing) of parents to the creation of a family by their adult children	80,4	6,6	13,0
2. Mandatory official registration of marriage	68,4	19,1	12,5
3. Creation of a family union between people of the same nationality	50,3	43,4	6,3
4. Mandatory presence of children in a family union	84,6	14,2	1,2
5. Obligatory male headship in the family	76,2	15,4	8,4

Comparative data in Table 1 show that for each of the presented traditional family customs, there is a high level of presence in the families of the respondents:

1. Mandatory presence of children in a family union (84.6%);
2. Mandatory consent (blessing) of parents to the creation of a family by their adult children (80.4%);
3. Mandatory headship of a man in the family (76.2%);
4. Mandatory official registration of marriage (68.4%);
5. Creation of a family union between people of the same nationality (50.3%).

Thus, despite modern dynamic changes in the social, information and communication sphere, according to the results of the survey, we see a high level of preservation of family traditions that regulate the normative field of decision-making in family life.

Distribution of answers to questions of the 3rd section – internationality of spiritual and moral values according to the youth community. Traditional family values are cultivated over long periods of time in the national culture of the people. According to the survey participants from among the student youth, the traditions of their parental family are characteristic of the majority of peoples living in the region (mentioned by 67.5% of the survey participants) and for the nationality to which the family members belong (27.4% of the survey participants). 5.1% of survey participants found it difficult to give a definite answer. 100% of survey participants are proud of the traditions of their people. Thus, we can say about the high level of the reproductive potential of traditional ethical and moral values among the students of the south of Russia. Despite the fact that traditions are influenced by modern conditions, we note the opinions of respondents about the high level of preservation of family traditions and ideas about normative values on the example of the stage of creating a family and the regulators of family life. They remain relevant and in many ways similar for different peoples living in the region and making up the regional and local community.

The results obtained during the empirical stage of the study do not contradict the ideas of other researchers. In the research of E. Ramos-Hidalgo, M. Orta-Pérez, M.A. Agustí we see the study of ethics and social responsibility issues through the prism of a family business [14]. Professional family traditions and ethical standards help to ensure the sustainability of the production sector in the socio-economic system of municipalities. Thus, the support of traditional family values is, among other things, a contribution to the sustainable development of the region [15].

For our study, there are also common key coincidences with the views of the authors L. Lu, T.T. Chang, S.F. Kao, which show the positive impact of traditional gender role values on the stability of production teams [16]. Thus, the preservation of traditional family values, ethical and moral guidelines help to enrich the socio-psychological environment of work collectives.

The Islamic concept of holistic ethical education, combined with modernity, according to B. Ben said, shows the possibility of adaptation, creativity, and development of intercultural experience [17]. Which in general contributes to the sustainable development of the region [18–20]. For our study, the aspect of the flexible applied nature of religious values in a multinational region, which is traditionally Stavropol Region and southern Russia, is relevant.

4 Conclusion

Consideration of research approaches to assessing the commitment of young people to traditional ethical and moral values, the impact on the stability of the regional community showed the high relevance and importance of this topic. Despite the differences that are obvious in regional, national and group segmentation, the positions and results of studies that are included in a brief thematic review, the positive impact of traditional spiritual and moral values on social and production processes is largely noted [21–23].

The results of the empirical part of the study show the high commitment of students of Stavropol State Agrarian University to traditional ethical and moral values. In the near future, young professionals will broadcast their commitment to the production teams of the agro-industrial sector of the region and the local community, which will contribute to sustainable regional development.

References

1. Sah, N.B., Banerjee, A., Malm, J., Rahman, A.: A good name is better than riches: family firms and working capital management. J. Behav. Exp. Financ. **33**, 100599 (2022)

2. Atman Uslu, N., YildizDurak, H.: Parental awareness and supervision to prevent cyberbullying: scale adaptation and a review in terms of demographic variables. Child Youth Serv. Rev. **133**, 106329 (2022)
3. Joseph, D., Belford, N., Lahiri-Roy, R.: Transnational daughters in Australia: caring remotely for ageing parents during COVID 19. Emot. Space Soc. **42**, 100864 (2022)
4. Gampe, A., Blaumeiser, J., Daum, M.M.: Children's group loyalty is related to parental in-group collectivism. Cogn. Dev. **61**, 101130 (2022)
5. Kim, J., Joo, S., Lee, K., Jun, H.: Family values in South Korean society: marital and sexual values in gender, age, and Protestantism contexts. J. Popul. Soc. Stud. **30**, 54–71 (2022)
6. Zawadzki, P., Teston, P., Lizote, S.A., Oro, I.M.: Organizational values: antecedents of well-being at work for rural successors. RAM. Revista de Administração Mackenzie **23**(1), 220118 (2022). https://doi.org/10.1590/1678-6971/eramg220118
7. Randerson, K.: Conceptualizing family business social responsibility. Technol. Forecast. Soc. Change **174**, 121225 (2022)
8. Cian, H., Dou, R., Castro, S., Palma-D'souza, E., Martinez, A.: Facilitating marginalized youths' identification with STEM through everyday science talk: the critical role of parental caregivers. Sci. Educ. **106**(1), 57–87 (2022). https://doi.org/10.1002/sce.21688
9. Dutton, H., Deane, K.L., Bullen, P.: Exploring the benefits and risks of mentor self-disclosure: relationship quality and ethics in youth mentoring. Kōtuitui **17**(1), 116–133 (2022). https://doi.org/10.1080/1177083X.2021.1951308
10. Sitthisopasakul, T., Kanjanawasee, S., Bhongmakapat, T.: Social returns on investment for doctor of philosophy program in educational measurement and evaluation. Kasetsart J. Soc. Sci. **42**(4), 878–885 (2021)
11. Liu, L., Zheng, F., Sheng, L., Hao, Y., Hu, J.: Reasoning talk at Chinese families' dinner table: across three generations and different communicative contexts. Front. Psychol. **12**, 763625 (2021)
12. Roberti, G.: Family wars and quarrels…about money and sharing|[Les guerresetguéguerres-familiales…à propos d'argentet de partage]. Cahiers Critiques de TherapieFamiliale et de Pratiques de Reseaux **67**(2), 127–139 (2021)
13. Lei, A., You, H., Luo, B., Ren, J.: The associations between infertility-related stress, family adaptability and family cohesion in infertile couples. Sci. Rep. **11**(1), 24220 (2021). https://doi.org/10.1038/s41598-021-03715-9
14. Ramos-Hidalgo, E., Orta-Pérez, M., Agustí, M.A.: Ethics and social responsibility in family firms. Research domain and future research trends from a bibliometric perspective. Sustainability (Switzerland) **13**(24), 14009 (2021)
15. Rivo-López, E., Villanueva-Villar, M., Novoa-Santos, S., Doval-Ruiz, M.I.: Does COVID-19 change CSR? A family business perspective. Sustainability (Switzerland) **13**(24), 13954 (2021)
16. Lu, L., Chang, T.-T., Kao, S.-F.: Sex and gender role orientation on the work-family interface: testing three-way interactions|[生理性別, 性別角色取向在職家互動的作用: 三階調節效果的檢驗]. NTU Manage. Rev. **31**(3), 1–46 (2021)
17. Bensaid, B.: An overview of Muslim spiritual parenting. Religions **12**(12), 1057 (2021)
18. Redon Pantoja, S., Vallejos Silva, N., AnguloRasco, J.F.: Education for citizenship: the meanings Chilean teachers convey in the neoliberal context. Sustainability (Switzerland) **13**(23), 13390 (2021)
19. Alisauskiene, M., Maslauskaite, A.: Religious identity and family practices in a post-communist society: the case of division of labor in childcare and housework. Religions **12**(12), 1040 (2021)
20. Gasanova, S.: Ethnic identity: peculiarities of interaction between family values and multi-ethnic student environment through the example of Dagestani students. Human. Soc. Sci. Commun. **8**(1), 276 (2021)

21. Willems, J.: Learning civil courage: a citizens' perspective. Educ. Res. **50**(9), 679–681 (2021). https://doi.org/10.3102/0013189X211044159
22. Singal, M., Batra, S.: The role of socioemotional wealth and entrepreneurial orientation in family-managed hotels. J. Hosp. Tour. Manage. **49**, 204–213 (2021). https://doi.org/10.1016/j.jhtm.2021.09.012
23. Walker, S.K., Lee, S.-K., Hong, S.: Workplace predictors of family educators' technology acceptance attitudes. Fam. Relat. **70**(5), 1626–1642 (2021). https://doi.org/10.1111/fare.12583

Rational Nutrition in Achieving the Sustainable Development Goals

Olga N. Kusakina, Angelina Ya Kazarova(✉), Angelika T. Aydinova, Yulia V. Rybasova, and Angelika R. Baicherova

Stavropol State Agrarian University, Stavropol, Russia
naukaekonom@mail.ru

Abstract. Sustainable development concept aims to achieve economic growth by ensuring social well-being and preserving the environment. This concept is often presented in the form of a trinity of economy, society and nature. At present, sustainable development is becoming increasingly relevant, since it is a reference point when developing a roadmap that covers almost all spheres of human activity. The article describes a plan for the implementation of the FoodNet roadmap for the period 2021–2035 and identifies the main directions of scientific activities in the field of designing new food products. The FoodNet market for the production of high-quality products is based on automation, robotization and intellectualization. The main goals and objectives of the FoodNet market are identified. Moreover, the possibility of integrating this market with hi-tech markets and the IT industry is substantiated. Finally, the prevailing directions of the FoodTech market in agriculture and personalized nutrition are indicated.

Keywords: Sustainable development · FoodNet · Agriculture · Personalized nutrition · Food market · Digital technologies

1 Introduction

The Sustainable Development Goals (SDGs) are a set of 17 interrelated goals developed on September 25, 2015 by the UN General Assembly as a plan to achieve a better and more sustainable future for all [12]:

1. Universal eradication of poverty in all its forms.
2. Eliminating hunger, ensuring food security and improving nutrition, promoting sustainable agricultural development.
3. Ensuring a healthy lifestyle and promoting well-being for everyone at any age.
4. Ensuring inclusive and equitable quality education and promoting lifelong learning opportunities for all.
5. Ensuring gender equality and empowering all women and girls.
6. Ensuring the availability and rational use of water resources and sanitation for all.
7. Ensuring access to affordable, reliable, sustainable and modern energy sources for all.

I. Samoylenko (Ed.): AMBT 2022, LNNS 582, pp. 91–100, 2023.
https://doi.org/10.1007/978-3-031-20803-4_10

8. Promote sustained, inclusive and sustainable economic growth, full and productive employment and decent work for all.
9. Building strong infrastructure, promoting inclusive and sustainable industrialization and innovation.
10. Reducing inequality within and between countries.
11. Ensuring openness, security, resilience and sustainability of cities and settlements.
12. Ensuring rational consumption and production models.
13. Taking urgent measures to combat climate change and its consequences.
14. Conservation and rational use of oceans, seas and marine resources for sustainable development.
15. Protection, restoration of terrestrial ecosystems and promotion of their rational use, rational forest management, combating desertification, stopping and reversing the process of land degradation and stopping the process of loss of biological diversity.
16. Promoting peaceful and open societies for sustainable development, ensuring access to justice for all and creating effective, accountable and participatory institutions at all levels.
17. Strengthening the means to achieve sustainable development and activating the mechanisms of the global partnership for sustainable development.

The objectives of SGD for the period up to 2030 ensure the balance of all three components of sustainable development: economic, social and environmental [14].

In order for the global community to achieve the Sustainable Development Goals (SDGs) by 2030, especially SDG 2 (eliminating hunger and malnutrition in all its forms and promoting sustainable agricultural development), food systems need to be transformed [10].

One of the acute socio-economic tasks today is to provide people with food. The food market is being formed, provided with intellectualization, automation and robotization of technological processes throughout the life cycle of products from production to consumption, as well as the development of biotechnologies [6].

In Russia, one of the main documents regulating nutrition issues is the Strategy for Improving the Quality of Food P[roducts in the Russian Federation until 2030. The purpose of the Strategy is to ensure the quality of food products as an essential component of health promotion, increasing the duration and improving the quality of life of the population of the Russian Federation, promoting and stimulating the growth of demand and supply for better food products, ensuring compliance with the rights of consumers to purchase quality products [16].

The Strategy goal will be achieved by solving the following tasks:

- regulatory framework development and improvement in order to protect consumer rights, development and/or creation of improvement of the methodological framework for assessing the compliance of food quality indicators;
- establishing food quality monitoring, improving state regulation and control in the field of food quality;
- creation of a unified information system for food quality traceability;
- assistance in the implementation of the food quality management system;

- creation of incentive mechanisms for producers to produce food products that meet quality criteria, as well as healthy food products;
- creation of conditions for the production of new generation products with specified quality characteristics;
- revival of food ingredients production in the Russian Federation;
- updating of the current standards for the content of food additives, flavoring substances, biologically active substances, residues of veterinary medicines and plant protection products in food products;
- priority development of scientific research in the field of human nutrition;
- ensuring the quality and safety of food products; studying the role of nutrition and the prevention of the most common non-communicable diseases; developing technologies for the production of high-quality food products;
- promotion of the principles of healthy nutrition [1].

Providing the population with food is one of the main goals of any Government and a prerequisite for national sovereignty. The problems of food and food safety and their impact on the health of the global population are complex, interrelated and cross-border. Therefore, international coordination is important not only to ensure the health of the global food system in the future, but also for the governments to fulfill their sovereign obligations to provide their own people with food [3].

A healthy diet is the key to a long life. The secret of longevity lies in a healthy and complete diet. Proper nutrition can affect our health, longevity and beauty. It assumes that many nutrients – proteins, carbohydrates, fats, water, minerals, vitamins – should be supplied to the body regularly, in the required amount and optimal ratios. A lack, as well as an excess of nutrients, firstly cause temporary inconveniences, then a source of disease development, a factor of premature aging and early death [5].

Rational nutrition is a balanced diet made up taking into account gender, age, health condition, lifestyle, nature of work and professional activity of a person, climate conditions of his residence. A properly formulated diet increases the body's ability to resist negative environmental factors, promotes the preservation of health, active longevity, fatigue resistance and high efficiency. The food we eat directly affects our vital functions, well-being and quality of life.

The purpose of the presented work is to analyze current problems, the implementation of which will allow us to move to a more sustainable path of economic development, create a fair system of distribution of benefits in society, significantly reduce the risks for future generations in the future existence and development. The realization of the goal set by us, in the ongoing scientific research, was achieved by performing the following tasks: (i) identification of opportunities for the introduction of digital technologies in order to ensure the sustainable development of the agri-food sector; (ii) consideration of the directions of development of the state economic policy in the field of ensuring food security of the country; (iii) study of the global Sustainable Development Goals; (iv) determining trends in the development of the Russian and global FoodNet market; (v) justification of the need and analysis of the basic principles of the introduction of personalized nutrition.

2 Informal Interpretation of the Problem

The goal of achieving food security is to guarantee sustainable supply of raw materials to processing enterprises, and continuous source of food to the population, which is not affected by external or internal adverse influences. It should not be vulnerable even in the event of rising prices, a shortage of foreign currency or an embargo on supplies abroad [5].

The most important conditions for achieving food security are:

- the physical availability of food for each person (the availability and supply in sufficient quantities);
- the economic ability of food purchasing by all social groups of the population, including the poor (solvency of consumer demand);
- consumption of high-quality products sufficient for rational nutrition [2].

Today, the educational and methodological platform of personalized nutrition is at the stage of formation.

The concept of optimal nutrition involves taking into account the characteristics of various social groups when developing their diets. This premise is of key importance for the further development of the theory of nutrition, since it was the first to formulate the need to differentiate diets depending on the characteristics of consumers. The development of the formulated thesis inevitably leads to the conclusion that the personalization of diets should be brought to the level of each individual consumer. The individuality of each person can be described in the first approximation by two categories of signs: physiological on the one hand and psychoemotional on the other [2].

It is necessary to develop products and diets that meet the physiological and psychoemotional needs of people, taking into account their gender, age, professional, geographical and national characteristics. The nutrition should be formed on the basis of special technologies that take into account the needs of social groups and individual consumers. In our opinion, social groups and individual consumers include:

- Schools and higher education institutions;
- Restaurants and cafes;
- Preschool educational institutions;
- Hospitals and sanatoriums;
- Pastry shops and bakeries [7].

Today, the concept of "healthy eating" is very popular, and everyone knows that it is better not to eat junk food. But few people know what exactly a diet should consist of so that it brings energy and health. The components of the human body, such as muscles, organs and bones, are made up of nutrients, and they can only be obtained from food. In particular, the human body requires six classes of nutrients (nutrients). These include [3]:

- water;
- carbohydrates;

– proteins;
– fats and fatty acids;
– vitamins;
– minerals.

A healthy diet is the consumption of whole foods without a high degree of processing (vegetables, cereals, nuts, eggs, fish). Such food is rich in fiber and protein, and contains fewer calories per serving than process food. A competent nutritionist takes into account many factors when adjusting nutrition: from a person's genetic predisposition to the current state of the psyche and stress level.

3 Formalization and Analysis of the Market

The market for the production of high-quality products, which is based on automation, robotization and intellectualization, is called FoodNet. The advantage of such market is that in this market that high-quality product is produced by the consumer at a lower price [9]. The new FoodNet market will be formed under the influence of the growing demands of consumers and the expansion of opportunities for the production of high-quality products based on the intellectualization, automation and robotization of technological processes throughout the cycle from production to consumption [8].

The FoodNet market can be divided into two directions: B2C and B2B. B2C is a market sector focused directly on the end consumers of manufactured products, including both general nutrition (traditional and food substitutes) and personalized (group and individual). B2B is a market sector focused on the organization of interaction between companies in the process of production and sale of food products by them.

The new food market Issues remain [12]:

– the use of various additives, such as flavor enhancers, preservatives;
– insufficient funding of scientific approaches, developments and innovations in agriculture;
– the problem of world hunger due to the limited resources and the low standard of living of the population.

Thus, the main goal of the FoodNet market is the process of creating high-tech products, taking into account human genetic characteristics. In this regard, the objectives of this market are:

1. Providing people with food;
2. Creation of a food market that provides high-quality products, which is based on automation, robotization and intellectualization;
3. Control throughout the entire process from the creation of the product to its consumption.

The history of personalized food as a business begins with the usual delivery of products from shops and restaurants and gradually moves from the general to the private, increasingly adapting to specific customers and their personal characteristics. Due to the emergence of new and cheaper existing technologies, we see recommendation systems that take into account the results of blood analysis, genetic analysis and microbiome analysis and select products for a specific person. The development of these diagnostic tests has become a big step towards the transformation of the personalized nutrition market.

In the last decade, "Personalized Nutrition" has been gaining popularity. This is due to the worldwide promotion of advanced countries in this area. However, there are a number of problems on the territory of the Russian Federation that hinder the development of this market. Such problems include:

1) This market is not in demand, since in some regions of the country this segment is little known.
2) The availability of an individual diet. Not everyone can afford it, because it is necessary to decipher the composition of the deoxyribonucleic acid (DNA) molecule test, which in Russia costs more than 8 thousand rubles, which not everyone can afford.
3) The complexity of treating a digital double of the consumer, taking into account genetic characteristics.

The FoodNet market will be closely integrated with hi-tech markets and the IT industry. The structure of the FoodTech market consists of:

- Innovative food products;
- Smart Packaging;
- IoT in agriculture;
- Biotechnologies;
- Personalized nutrition;
- Sustainable Packaging.

The structure of the FoodTech market in the main directions is presented in Fig. 1.

According to the values of Fig. 1, it is possible to identify the prevailing trends in the FoodTech market – Sustainable Packaging, IoT technologies in agriculture and personalized nutrition. There is a tendency for the following segments of the FoodTech market to increase, namely: "innovative food products" by 0,1%, "biotechnology" by 1,7%, "personalized nutrition" by 5.9%.

Let us highlight several global vectors of FoodTech market development:

- Delivery from restaurants is the most ambitious direction. This is the marketplace model (in many ways identical to Booking.com, only works in the field of food orders from restaurants), that is, companies work as large-scale order aggregators that help people make a choice among many restaurants and eventually transfer the order directly to the selected one. At the same time, companies often provide the delivery service themselves. Such a format is represented by both multinational players, such as Delivery Hero and FoodPanda, and local, but very large ones, for example, Chinese Ele.me.

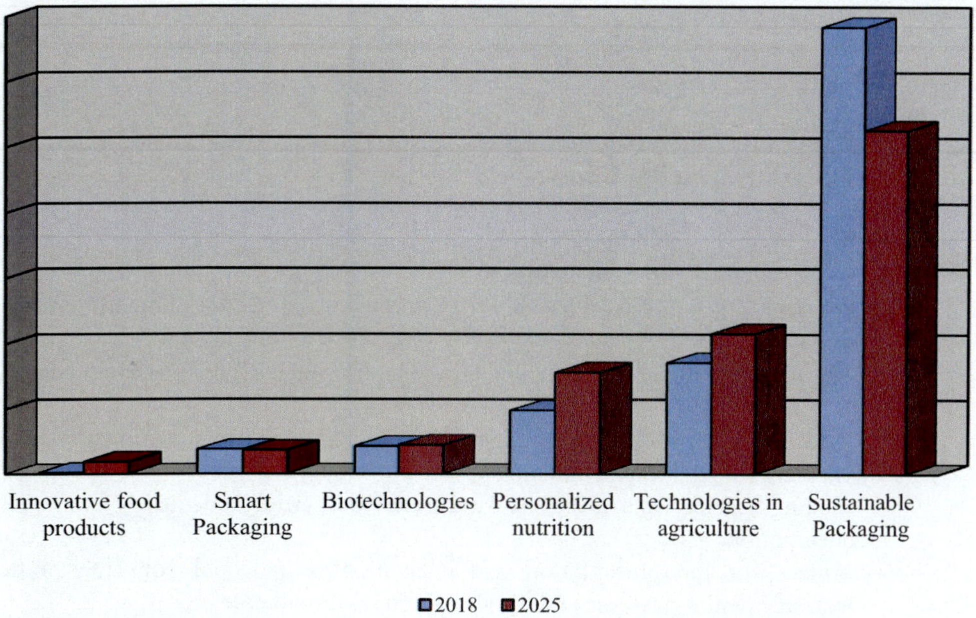

Fig. 1. The structure of the FoodTech market by main directions [11]

- Ready meals on request. Different models can be used here, but in the end it is the delivery of ready meals either on schedule or by request. The choice of dishes is usually limited, but the menu is different for each day. The most prominent global representatives in this niche are Sprig, Munchery and Maple. Here, with some stretch, you can add services for the delivery of food to offices.
- Sets of products for cooking by subscription. Services for those who prefer to cook independently according to the provided recipes and a basket of products. It is convenient that everything necessary to prepare the selected dish in the right amount is delivered to the client. Such services are provided by such companies as HelloFresh and Blue Apron and in the Russian market we can mention the company "Food Party".
- Delivery of farm products. For those who want to receive only fresh products "straight from the garden", services like FarmDrop and Farmigo provide the opportunity to order directly from the farms.
- New approaches in food and beverages. It is not yet clear what their prospects are and how a new approach to food will be perceived, but it is impossible not to mention such bright companies as Impossible Foods and Juicero. Their proposals radically change traditional ideas about the culture of nutrition. Almost all of the above-mentioned models are represented on the Russian market at different stages of development. Any local market is influenced by national peculiarities of food consumption, popular restaurants and cooking. For example, in the Middle East and Asia as a whole, it is customary to order food delivery at home. Russia in this sense also has its own peculiarities, but in our case it rather slows down the development of "classical" models. For example, the culture of visiting restaurants has emerged relatively recently in Russia. So far, in most cases, going to a restaurant is considered a special event, and

not just a way of consuming food. Therefore, food delivery services from restaurants are perceived as an additional and extent even a festive occasion.

4 Modern Technologies Used in Providing the WORLd's Population with Food

Nutrition strategies should not only be aimed at guaranteeing food safety (i.e. sufficient nutrition), but also ensure the consumption of an adequate amount of healthy and high-quality food. The analysis of trends and patterns of food consumption around the world is becoming relevant.

The FoodNet roadmap implementation plan for the period 2021–2035 considers:

- Smart devices for testing and express determination of the nutritional status of the consumer by vital trace elements;
- Tracking services and a digital database of properties and quality of food before consumption;
- Services for the presentation and evaluation of personal consumer indicators, electronic health passport and a set of personal food recipes;
- Services for automating decision-making on the selection of the necessary personalized diet based on digital data of the consumer and the product;
- Technologies and equipment for creating mass production of personalized food replacement, which is based on technologies of personalization of food additives;
- Cryobank of personal microbiota with the ability to store material throughout the life of the consumer and service for the preparation of personalized food with the addition of personal microbiota;
- Technologies and equipment for the production of personalized food products at home environ for target groups nutritional, taste and other needs based on additive technologies [15].

The solution to the world hunger problem is to eliminate irrational usage of products. Reducing food losses during processing and storage can cut the forecasted shortage of food by 20% by 2050 [4]. There is a need to create and implement new technologies to solve the problems of rapidly growing water shortage and reduce losses in processing and storage.

The longer the pandemic lasts, the more advantages there are for companies that have already developed their technologies aimed at producing individual nutrition. This does not mean that the FoodNet market should oust out traditional markets, such as: food and agricultural. This market will take a leading position among them. Thus, the main directions that should lead to an increase in the market will be: biotechnology, innovative food products, and the Internet of Things (IoT). The main prevailing directions of the FoodTech market will be: Sustainable Packaging (eco-friendly, recyclable packaging in the Food&Beverage segment), IoT in agriculture and personalized nutrition.

In developed countries, there is a request for a healthy diet. Ratings of the best nutrition apps are compiled that help to follow a healthy diet, count calories, motivate and even help to make purchases [9].

The main directions of scientific and scientific-practical activity in the field of designing new food products are:

- development of recipes and technologies of mass-consumption food products, functional and specialized food products;
- development of formulations and technologies of biologically active food additives, vitamin and mineral premix-fortifiers and technological mixtures intended for food enrichment;
- study of physico-chemical and organoleptic parameters of functional and specialized food products;
- study of the content and safety of micronutrients in raw food components and fortified foods during their production and storage;
- scientific and analytical examination of biologically active additives and food products enriched with vitamins, minerals and other functional food ingredients;
- creation of software products for automated calculation of formulations taking into account fluctuations in the composition of raw materials and optimization of the composition of multicomponent prescription mixtures [13].

Automation of the recipe design process will allow employees of food enterprises to respond promptly to changes in the properties and types of raw ingredients, changes in consumer preferences, and create products with a predetermined chemical composition, nutritional value and functional orientation.

Optimal solutions to these problems in the design of food products can be achieved with the help of their formalized mathematical descriptions – mathematical models that reflect in an analytical form many functional relationships between technological, economic and other parameters of raw ingredients, the required characteristics of finished products (objective function) and a number of restrictions arising from the requirements of regulatory documentation. Thus, mathematical and simulation modeling becomes one of the necessary tools for the technologist to solve problems of optimizing the complex properties of a food product – organoleptic, physico-chemical, microbiological.

5 Conclusion

The paper considers the goals in the field of sustainable development of rural areas. Each goal is aimed at solving a specific global problem. Their implementation will make it possible to move to a more sustainable path of economic development, create a fair system of distribution of benefits in society, significantly reduce the risks for future generations in the future existence and development.

The world has all the necessary resources to produce enough food for the entire population of the planet. Modern food problems are associated with irrational methods of harvesting, food losses, inadequate distribution. The food and agriculture industries offer a variety of development strategies that will eliminate hunger in the long term and ensure universal food security.

Agriculture is the largest employer in the world, providing livelihood for 40% of the world's population. This is the biggest source of income and jobs.

In our opinion, the best way to save people from hunger is to invest in the agricultural sector. Major investments in sustainable land use and land restoration will contribute to improving food security, strengthening livelihoods and promoting adaptation to climate change.

References

1. On the approval of the Strategy for improving the quality of food products until 2030. Government of the Russian Federation. http://government.ru/docs/23604/
2. Anishchenko, A.: "Smart" agriculture as a promising vector of growth of the agricultural sector of the Russian economy. Food Policy and Security, p. 106. Moscow (2019)
3. BBSRC: BBSRC research in food, nutrition and health strategic framework: 2015–2020. http://www.bbsrc.ac.uk/documents//1503-fnh-strategic-framework/
4. WHO: Diet, nutrition and the prevention of chronic diseases. http://www.fao.org/docrep/005/AC911e/ac911e05.htm
5. Golden, C., et al.: Nutrition: fall in fish catch threatens human health. http://www.nature.com/news/nutrition-fall-in-fish-catchthreatens-human-health-1.20074
6. Koniordos, M., Matvienko-Bilyaeva, G., Strapchuk, S.: Strategic Scenario of an Open Source of Sustainable Development for the Food System, p. 56 (2017)
7. Michael, E.: Lean, principles of human nutrition. Medicine (United States of America) **3**, 140–144 (2019)
8. Chernukhi, I., Ivanova, V., Sidorenko, Y.: Personalized Nutrition: Designing Products and Diets: A Textbook, p. 462. Moscow (2020)
9. Shaulina, L., Korsun, L.: Quality Control and Safety of Food Products and Food Raw Materials: Textbook, p. 111. IGU, Irkutsk (2011)
10. Mainstreaming the 2030 Agenda for Sustainable Development. Reference Guide to UN Country Teams
11. Healthline Media: The Best Nutrition Apps of 2016. http://www.healthline.com/health/-food-nutrition/top-iphone-android-apps
12. Official website of the Government of the Russian Federation: The program "Digital Economy of the Russian Federation". http://government.ru/rugovclassifier-/614/events/
13. Tutelyan, V.: Scientific approaches to the assessment of consumption levels of biologically active substances. http://vrn.drnona-net.ru/conferences/medconference5/1444.html
14. United Nations Development Group: https://undg.org/main/undg_document-/mainstreamingthe-2030-agenda-for-sustainable-development-interim-reference-guide-to-uncts/ Feb (2016)
15. Universal Sustainable Development Goals. Understanding the Transformational Challenge for Developed Countries. Report of a Study by Stakeholder Forum (2015)
16. Vorobyova, N.V., Gracheva, D.O., Bannikova, N.V., Telnova N.N., Cheremnich, M.B.: Perspective directions of the agri-food market development in the system of bilateral foreign trade relations between Azerbaijan and the Stavropol Territory. In: IOP Conf. Series: Earth and Environmental Science. IOP Publishing Ltd., Russia (2021)

Training of Veterinary and Sanitary Professionals for the Sector and Region Sustainable Development

Valentina Ivashova(✉), Olga Chudnova, Roza Chvalun, Irina Makhova, and Natalya Kizilova

Stavropol State Agrarian University, Stavropol, Russia
vivashov@mail.ru

Abstract. The article describes the theoretical aspects of studying the request of the professional community for the personal and professional qualities of graduates of the educational Master's program "Veterinary and Sanitary Expertise". A review of research strategies and results in the field of ensuring the safety of agricultural products showed the need to supplement the profile of specialists with digital competencies, competencies in statistics, microbiological research tools and attention to personal safety in working with biologically hazardous objects. An empirical study was carried out with the participation of representatives of the professional community of enterprises and organizations of the Stavropol Territory. Based on the expert survey, the key attributes of the program graduate's profile are highlighted, the most significant for successful work in organizations that carry out veterinary and sanitary expertise and ensure the biological and environmental safety of products of animal and plant origin.

Keywords: Professional qualities · Profile of specialist · Professional competencies · Professional community

1 Introduction

The training of a modern specialist for any branch of the economy cannot be separated from the real needs of employers. The dynamic change in production processes generates a request for new competencies of specialists necessary for successful work and associated with the prospects for the development of enterprises and organizations. Of course, each industry has its own specifics and key development trends. So the enterprises of the agro-industrial complex have a common basis for the objects of professional activity, which, as a rule, are living biological systems: plants and animals. Any biological system develops according to certain life cycles, closely related to natural and climatic conditions, which determine the risk factors for the agro-industrial complex.

The conducted informative review of publications on the issues of veterinary and sanitary safety of livestock products confirmed the relevance of the issue under study. Modern training of graduates of the Master's program "Veterinary and Sanitary Expertise" requires constant updating of the competence profile, depending on the production

I. Samoylenko (Ed.): AMBT 2022, LNNS 582, pp. 101–110, 2023.
https://doi.org/10.1007/978-3-031-20803-4_11

requirements. A review of research strategies and results in the field of safety of livestock products showed the need to supplement the profile of specialists with digital competencies, competencies in statistics, microbiological research tools and ensuring personal safety in working with biologically hazardous objects. This information became the basis for the development of tools for the empirical part of the study of the actual competence profile of graduates of the Master's program "Veterinary and Sanitary Expertise".

The authors of the study A. Shemyakin, Y. Lyashchuk, A. Martynushkin, E. Strokova, and A. Krasnikov assess the level of biological risks of the activities of enterprises in the agro-industrial complex at the regional level [1]. The results of the study show that biological risks associated with the use of low-quality agricultural products are the main ones due to the fact that they pose a threat to human life and health. Veterinary and sanitary professionals should be experts in infectious diseases, epizootology, epidemiology and microbiology, contributing to the identification of threats posed by agricultural products. Expert work is associated with a high level of social and professional responsibility and this should be reflected in the design of the competence profile of the future specialist of the veterinary and sanitary service.

Risks and threats of a sanitary and epidemiological nature for the population are also objects of wild nature that affect the ecosystem parameters of the regions. The authors of the study M.N. Sidorov, E.P. Tomashevskaya, A.N. Maksimova show the need to determine the quality of meat of wild ducks, the scale of capture of which is increasing in the northern region – the Republic of Sakha (Yakutia) [2]. Such work is important for assessing possible threats to human health. The accuracy of such an assessment determines the meaningful component of informing the population and, accordingly, requires the use of precise methods and strategies for data verification. In the course of constructing the professional block of competencies of the educational Master's program "Veterinary and Sanitary Expertise", the ability to use the tools of microbiological research and draw conclusions that provide an information basis for making management decisions and informing the public should be distinguished as structural units.

The diversity of competencies of graduates of the Master's program is due to the fact that their employment can be carried out both in organizations that control the sanitary and epidemiological situation, and in enterprises of the production sector, for example, poultry farms. The authors of the study are N.A. Zhuravel, A.V. Miftakhutdinov, V.V. Zhuravel show that highly professional veterinary and sanitary control and bacteriological monitoring of products at a poultry farm reduces the risks of producing unsafe poultry meat [3]. In general, this has a positive effect on the efficiency of the enterprise and increases the importance of the functionality of a specialist – veterinary sanitary expert.

Intensive livestock rising is necessary to solve the problem of providing the population with high-quality meat products. One of the directions of intensification of beef cattle breeding is the use of highly productive breeds of livestock and breeds that are atypical for a particular area. The authors of the article D. Zamaratskii, M. Simakova, G. Koltun, V. Podvalova on the example of Primorsky Krai show how natural focal parasitic diseases of animals sharply reduce the efficiency of beef cattle breeding [4]. Organization of veterinary and sanitary protection of livestock and timely measures to prevent invasive diseases of animals on farms, as well as careful monitoring of slaughter products with subsequent systemic conclusions at the level of meat producers and territorial veterinary and sanitary control will provide the population with safe conditions for consumption. Thus, graduates of the Master's program "Veterinary and Sanitary Expertise" carry out a humanitarian mission - preserving the health and safety of the population.

The work of the veterinary sanitary expert plays an important role in organizing a barrier to the penetration of diseases dangerous to humans, the nature of which is determined by poor-quality meat products. With insufficient attention to the presence of zoonotic agents in the results of microbiological studies of meat, diseases dangerous to humans can develop. Public health protection includes a thorough inspection of meat for a microbiological assessment of threats. The authors of the article A. Didkowska, P. Zmuda, E. Kwiecień, B. Orłowska, K. Anusz, using the example of postmortem examination of slaughtered sheep, draw systemic conclusions about the possibility of spreading consumer diseases through contaminated lamb meat [5]. In the course of describing the experimental work, we can highlight important professional competencies for graduates of the Master's program "Veterinary and Sanitary Expertise": knowledge of the methodological and methodological tools of experimental work, including the statistical characteristics of sampling, analytical system conclusions, microbiological examinations, systemic knowledge of a biological object and its life cycle.

2 Materials and Methods

An expert survey of employers in the person of managers, chief and leading specialists in the field of training 36.04.01 – Veterinary and Sanitary Expertise, profile "Biological and Environmental Safety of Products of Animal and Plant Origin" was carried out by the method of electronic questionnaire in Google Form in May-June 2021. In total, 24 people from 17 specialized organizations and enterprises took part in the survey:

The purpose of the expert survey was to determine the current competence profile of the graduate of the Master's educational program "Biological and Environmental Safety of Products of Animal and Plant Origin".

Objectives: to consider the expectations of the professional expert community regarding the universal, general professional and professional competencies of graduates; determine the working conditions at enterprises and organizations responsible for the biological and environmental safety of products of animal and plant origin, and the possibility of hiring graduates of this profile.

3 Results and Discussion

General views of the expert professional community about the qualities of specialists hired are presented in the following Table 1.

The ranking of the significant qualities of the specialists recruited showed that in the first place, with a large margin from other options, are good knowledge in their field, which, according to experts, is the main attribute of a young specialist. Next in importance are good personal potential and the presence of a diploma of graduation from higher education. The option 'work experience', although it scored 55.0% of the

Table 1. Distribution of answers to the question about the importance of personal and professional qualities of specialists employed (%)

List of qualities	Expert answers, %	Rank 1	Alumni responses, %	Rank 2
1. Good personal potential	60,0	2,5	42,0	3
2. Good knowledge in professional field	95,0	1	85,5	1
3. Having a diploma of graduation from higher education	60,0	2,5	23,2	5
4. Availability of additional education, several qualifications, courses	10,0	6	11,6	7
5. Work experience	55,0	4	69,6	2
6. Willingness to work regardless of wages	20,0	5	30,4	4
7. Ability to present oneself	5,0	7	18,8	6
8. Nice appearance	–	8	8,7	8
9. Having good health	–	8	5,8	10
10. Demographic factors (gender, age, nationality)	–	8	7,2	9
11. Other qualities (to specify)	–	7	–	–

answers, is only in 4th place in terms of importance. In general, the listed characteristics make up a significant attributive portrait of a specialist who is hired by employers.

It cannot be said that these views of the experts fully coincide with the vision of the graduates of the educational program. Just like experts, graduates give first place to good knowledge in the professional field (85.0% of survey participants); in second place is work experience (69.7% of survey participants), while the expert community ranked this attribute only in 4th place with a value of 55.0%; in third place, both graduates and experts put good personal potential, but experts attach much more importance to this attribute than graduates: 60.0% compared to 42.0%.

Comparison of attributive characteristics in the answers of graduates and experts shows the need for the formation and high importance of universal competencies that develop the personal potential of future employees.

Table 2 presents data characterizing the conditions of an employment contract for graduates of an educational program that employers can provide at the workplace.

Table 2. Distribution of answers to the question about the conditions that are provided to employees – qualified specialists with higher education (%)

List of conditions	%	Rank
1. Timely payment of wages	90,0	2
2. Full social package (payment of vacation pay, etc.)	95,0	1
3. Training at the expense of the organization, obtaining additional courses	70,0	3
4. Ability to travel, relocation, business trips	–	7
5. Comfort of the workplace	5,0	6
6. Good relationships in the team, with the leadership	20,0	4
7. Opportunity to grow up the career ladder	20,0	4
8. Opportunity to travel to sanatoriums, dispensaries on benefits	–	7
9. Provision of transport	10,0	5
10. Other conditions (to specify)	–	7
11. Difficult to answer	–	7
12. None of the above conditions	–	7

The list of conditions is headed by such as a full social package (95.0%), timely payment of wages (90.0%), training at the expense of the organization (70.0%). To assess the possibilities for the development of additional education at the university, important information is the willingness of enterprises and organizations in the educational program of "Veterinary and Sanitary Expertise" to pay for advanced training of their employees (Table 3).

Table 3. Distribution of answers to the question: "Do you focus on professional standards in your work, where the labor functions of specialists are spelled out?" (%)

Answer options	%
1. Yes, we mainly focus on prof. standards	70,0
2. No, we have our own approaches to the organization	10,0
3. I can't say for sure	20,0
4. Other options	–

In their activities, the survey participants and the number of the expert community are mainly guided by professional standards – noted by 70.0% of respondents. 10.0% of survey participants said about other approaches and 20.0% were unable to give a definite answer.

Thus, we can say that professional standards play an important role in the definition and distribution of labor functions of specialists in the educational program "Veterinary and Sanitary Expertise" (Table 4).

Table 4. Distribution of answers to the question: "Do you hire young specialists to your organization (enterprise) after graduation?" (%)

Answer options	%
1. Yes	100,0
2. No	–

In the course of the survey, representatives of the professional community were unanimous in their answer to the question: "Do you hire young specialists to your organization (enterprise) after graduation?" In 100% of cases, young specialists are hired.

Moreover, a probationary period for young specialists was introduced only in half of the organizations. The data is presented in Table 5. 15% of the experts said that there was no probationary period. And another 35.0% noted that the probationary period is assigned individually, depending on the characteristics of the candidate, which once again emphasizes the need to build personal potential, and not only knowledge in the professional activity of graduates during the period of study at the university.

Table 5. Distribution of answers to the question: "Is there a trial period when applying for a job in your organization (enterprise)?" (%)

Answer options	%
1. Yes	50,0
2. No	15,0
3. Decisions are made on an individual basis	35,0

55.0% of experts note that they would like to hire a specialist with experience, but not older than 40 years. But it should be noted that the share of those experts who indicated that they would have made a choice in favor of young specialists is quite large – 30.0% of the number of survey participants. The data is presented in Table 6.

In the course of the survey, the experts named the labor functions that a university graduate should be able to perform, working as a specialist in an organization (answers are given in the authors' transcription): an epizootologist, first of all, according to the job description, must treat and prevent animal diseases; organize veterinary preventive measures; perform the functions of a veterinarian; apply knowledge and skills in practice; be able to organize work for high-quality performance of labor duties; be competent in veterinary matters; the specialist must ensure the prevention of the spread of diseases and their pathogens from foci or disadvantaged regions; to prevent the introduction of diseases from disadvantaged regions to prosperous ones; creation of opportunities for the safe movement of animals and other goods controlled by the state veterinary service between the constituent entities of the Russian Federation; have responsibility and diligence; perform their job duties efficiently; know and be able to carry out VSE

Table 6. Distribution of answers to the question: "Who would you rather hire: a young specialist or an experienced specialist?" (%)

Answer options	%
1. Young specialist	30,0
2. A specialist with experience, but not older than 40 years	55,0
3. A specialist with experience, the rest does not matter	5,0
4. Other options	–
5. Difficult to answer	10,0

products, operate in the "Mercury" system, understand what he is doing and want to work.

In general, attention is drawn to the high proportion of formulations that are associated with the personal potential of a young specialist, they account for about a third of the entire list.

At the national level, countries are working on an expert assessment of the biological hazard of natural objects, which is the basis for updating the requirements in veterinary and sanitary-epidemiological legislation. A separate task is to study old burials of animals that died during dangerous epidemics, including anthrax. The authors of the article are Z.F. Dugarzhapova, M.V. Chesnokova, T.A. Ivanova, S.A. Kosilko, S.V. Balakhonov offer an improved methodology for examining anthrax burial grounds [6]. According to a group of scientists, a comprehensive epizootological and epidemiological survey of burial grounds makes it possible to regional map the threats of the spread of epidemics, which is important for large territories of the country and increasing anthropogenic loads on ecosystems that have developed over decades. For our study of the competence profile of future veterinary experts, it is relevant to understand the place of the skills of conducting complex epizootological and epidemiological surveys, which pose a danger not only to the population, but also to the researchers themselves. And also the readiness to summarize the results of such surveys, which will be further used in the form of information bases (digital competencies for working with data) and proposals for improving the legal regulation of veterinary and sanitary safety (competencies associated with legislative initiative).

The expansion of the scale and accuracy of research on animal diseases reveals new levels of understanding of their relationship to disease and human hazards. The authors of the article P.A. Brown, C. Courtillon, E.A.W.S. Weerts, M.H. Verheije, N. Eterradossi give the results of the dynamics of production and transmission of infection by the example of experimental infection of turkeys [7]. The authors emphasize that the rate of spread of viral infection in birds shows the importance of sanitary prevention on farms and strict veterinary and sanitary control. Accordingly, qualified specialists with professional competencies are needed at the request of employers who are ready to provide a set of preventive measures and monitoring of the veterinary and sanitary situation at enterprises in the agricultural sector [8–10].

The relevance of studying the request of the professional community in order to update the competence profile of graduates of the Master's program "Veterinary and Sanitary Expertise" is confirmed by the results of a number of studies. Authors V.I. Khizgiyaev, N.I. Briko, V.R.. Kuchma, T.I. Gordeeva, N.A. Volkova consider the professional standard of a specialist in the field of preventive medicine as the basis for training and activities in the field of medical prevention [11]. The development of an updated version of the professional standard was attended by representatives of enterprises and organizations that hire graduates of programs of treatment and prevention. Thus, employers are directly involved in the formulation of the functionality required by modern production. The procedures for securing the expert opinion of the professional community in specific regulations are important in understanding and optimizing algorithms for new institutional contours of food sanitary safety, taking into account the variety of forms of livestock production and the current epidemiological situation [12–17].

The carried out substantive analysis of publications on the issues of veterinary and sanitary safety of livestock products confirmed the relevance of the issue under study. Modern training of graduates of the Master's program "Veterinary and Sanitary Expertise" requires constant updating of the competence profile, depending on the production requirements. A review of research strategies and results in the field of safety of livestock products showed the need to supplement the profile of specialists with digital competencies, competencies in statistics, microbiological research tools and ensuring personal safety in working with biologically hazardous objects. This information can serve as the basis for the development of tools for the empirical part of the study of the current competence profile of graduates of the Master's program "Veterinary and Sanitary Expertise".

4 Conclusion

The analysis of the results of an expert survey conducted at enterprises and organizations of the Stavropol Territory, whose activities are related to ensuring the veterinary and sanitary safety of agricultural products, made it possible to draw a number of conclusions:

1. Ranking of the significant, according to experts, qualities of specialists hired showed that in the first place, with a large margin from other options, are 'good knowledge', which is the main attribute of a young specialist. Next in importance are good personal potential and the presence of a diploma of graduation from higher education. The listed characteristics make up the actual attributive portrait of a specialist who is hired by employers.
2. These views of the experts do not fully coincide with the vision of the graduates of the educational program. Just like experts, graduates give first place to good knowledge in the subject area (85.0% of survey participants); in second place is work experience (69.7% of survey participants), while the expert community ranked this attribute only in 4th place with a value of 55.0%; in third place, both graduates and experts put good personal potential, but experts attach much more importance to this attribute than graduates: 60.0% compared to 42.0%.

Comparison of attributive characteristics in the answers of graduates and experts shows the need for the formation and high importance of universal competencies that develop the personal potential of future employees.

References

1. Shemyakin, A., Lyashchuk, Y., Martynushkin, A., Ivanishchev, K., Strokova, E., Krasnikov, A.: Analysis and assessment of the level of biological risks of activities of enterprises of the agro-industrial complex at the regional level. E3S Web of Conf. **258**, 06057 (2021). https://doi.org/10.1051/e3sconf/202125806057
2. Sidorov, M.N., Tomashevskaya, E.P., Maksimova, A.N.: Veterinary-sanitary examination of migratory bird carcasses when stored in a glacier under permafrost conditions. IOP Conf Ser.: Earth Environ. Sci. **666**(2), 022057 (2021)
3. Zhuravel, N.A., Miftakhutdinov, A.V., Zhuravel, V.V.: Identification and risk assessment in determining the effectiveness of production veterinary and sanitary control in poultry farms. IOP Conf. Ser.: Earth Environ. Sci **624**(1), 012180 (2021)
4. Zamaratskii, D., Simakova, M., Koltun, G., Podvalova, V.: Veterinary and sanitary examination of yak slaughter products in case of invasive diseases in the conditions of the Primorsky Territory. E3S Web of Conf. **203**, 01031 (2020). https://doi.org/10.1051/e3sconf/202020301031
5. Didkowska, A., Zmuda, P., Kwiecień, E., Orłowska, B., Anusz, K.: Microbiological assessment of sheep lymph nodes with lymphadenitis found during post-mortem examination of slaughtered sheep: Implications for veterinary-sanitary meat control. Acta Vet. Scand. **62**(1), 48 (2020). https://doi.org/10.1186/s13028-020-00547-x
6. Dugarzhapova, Z.F., Chesnokova, M.V., Ivanova, T.A., Kosilko, S.A., Balakhonov, S.V.: Improvement of methodical approaches to investigation of anthrax burials and animal burial sites. Problemy Osobo Opasnykh Infektsii. **2019**(4), 41–47 (2019)
7. Brown, P.A., et al.: Transmission Kinetics and histopathology induced by European Turkey Coronavirus during experimental infection of specific pathogen free turkeys. Transbound. Emerg. Dis. **66**(1), 234–242 (2019). https://doi.org/10.1111/tbed.13006
8. Satyaprakash, K., Khan, W.A., Chaudhari, S.P., Shinde, S.V., Kurkure, N.V., Kolte, S.W.: Pathological and molecular identification of porcine cysticercosis in Maharashtra, India. Acta Parasitol. **63**(4), 784–790 (2018). https://doi.org/10.1515/ap-2018-0094
9. Allain, V., Salines, M., Le Bouquin, S., Magras, C.: Designing an innovative warning system to support risk-based meat inspection in poultry slaughterhouses. Food Control **89**, 177–186 (2018)
10. Suleiman, T.S., Karimuribo, E.D., Mdegela, R.H.: Prevalence of bovine subclinical mastitis and antibiotic susceptibility patterns of major mastitis pathogens isolated in Unguja island of Zanzibar, Tanzania. Trop. Anim. Health Prod. **50**(2), 259–266 (2018)
11. Khizgiyaev, V.I., Briko, N.I., Kuchma, V.R., Gordeeva, T.I., Volkova, N.A.: Professional standard of specialist in the field of medical and prophylactic case as the basis of preparation and activity in the field of medical prevention. Gigienai Sanitariya **96**(4), 382–387 (2017)
12. Bavyko, A.: Nutritional value and safety of aquaculture products in Ukraine. J. Hyg. Eng. Des. **21**, 41–49 (2017)
13. Poglayen, G., et al.: Retrospective study on cystic echinococcosis in cattle of Italy. J. Infect. Dev. Ctries. **11**(09), 719–726 (2017). https://doi.org/10.3855/jidc.9433
14. Baygazanov, A., Bashkina, Y., Derbyshev, K., Nurkenova, M.: Seasonally age dynamic of ascospherosis disease of bees and its progress in East Kazakhstan. Res. J. Pharm., Biol. Chem. Sci. **7**(6), 1743–1749 (2016)

15. Bersano, J.G., et al.: Demodex phylloides infection in swine reared in a peri-urban family farm located on the outskirts of the Metropolitan Region of São Paulo, Brazil. Vet. Parasitol. **230**, 67–73 (2016). https://doi.org/10.1016/j.vetpar.2016.06.014
16. Seregin, I.G., Nikitchenko, V.E., Plyushchikov, V.G., Nikitchenko, D.V., Kubatbekov, T.S.: Veterinary-sanitary examination of domestic and wild duck meat at sarcocystosis. Information (Japan) **19**(10), 4935–4942 (2016)
17. Nikitin, A.I., et al.: Veterinary and sanitary examination of meat of the animals subjected to physical and biological effects. Res. J. Pharm. Biol. Chem. Sci. **7**(4), 2207–2213 (2016)

The Increase of Labour Efficiency and Innovative Regionalization in Russia

Natalia V. Bannikova(✉), Tatiana N. Kostyuchenko, Natalia N. Telnova, Angelika R. Baicherova, and Marina B. Cheremnykh

Stavropol State Agrarian University, Stavropol, Russia
nbannikova@mail.ru

Abstract. In full accordance with the goals of sustainable development determined by the United Nations until 2030 and the national development goals of the Russian Federation, it is necessary to ensure superior, as compared to worldwide average, GDP (gross domestic product) growth, its value calculated per 1 employee being the index of social labor productivity. The goal of the research is to determine the interrelation between labour productivity, innovative activity of organizations and the level of small business development with the use of correlation-and-regression analysis. The specific feature of the present research is the investigation of interregional differences of the effects under consideration by performing the comparison of three macro-regions' data. The closeness of interrelation of the investigated factors for Russia as a whole is estimated as average. At the same time, the impact of small business development upon changes in labour productivity level is negative. Much more significant impact of innovative activity upon the dynamics of labour productivity is registered in central and north-west Russia, where main industry knowledge-intensive expertise centers are located. There is almost no interrelation between the investigated factors in southern regions. This leads to the necessity to create the specific ecosystem of innovative development with the use of «catching-up development» and polycentricity principles.

Keywords: Innovations · Labour productivity · Macro-region · Small business · Ecosystem of innovative development

1 Introduction

For the beginning of 2021 Russia was in the 67th place among the world countries in terms of standard of living, lagging behind the leader (Switzerland), in particular, by 3.3 times in consumer purchasing power and by almost 4.0 times in the cost of living [22]. Russian population steadily decreases. Population age distribution is getting worse, because the share of working-age people decreases, while the number of pension-age people increases. Under these circumstances, it is possible to ensure the growth of people's standard of living only on the basis of labour performance increase. However, also in its growth rate of labour productivity Russia is far behind the leader in average

I. Samoylenko (Ed.): AMBT 2022, LNNS 582, pp. 111–123, 2023.
https://doi.org/10.1007/978-3-031-20803-4_12

annual growth – China – by 5.0 times, and behind the leader in the approached level of this index – the USA – by 2.3 times [15].

That is why one of the main objectives in the Decree of the President of Russia No. 474 "On the national development goals of the Russian Federation for the period up to 2030" is "decent and effective work and successful business activity". The achievement of this objective is characterized by ensuring the growth of country's gross domestic product (GDP) higher than worldwide average and real growth of capital investments. It is not possible to achieve this goal without the growth of organizations' innovative activity throughout all subjects of the Russian Federation. This has been proved by the investigations of such science classicists as Y. Schumpeter, R. Solow, E. Denison and S. Kuznetz, P. Romer [25].

The most significant influence of innovative factors upon the growth of country's economy is also confirmed in the investigations of many contemporary academic economists: Ya. Nikonova, E. Basovskaya, R. Khanna & C. Sharma, S. Kurt & Ü. Kurt [4, 13, 14, 18]. At the same time, this issue cannot be considered as definitely resolved, because there are still ongoing disputes regarding the methods for the evaluation of specific results of innovative processes in short-term and long-term periods, as well as the ways of transforming new knowledge and technologies into the accelerators of economic development at macroeconomic level.

The problem of interrelation between the innovative activity and the indicators, which describe labour productivity, is still paid great attention. The authors are interested in different aspects of these complicated and many-sided cause-and-effect relations between innovations and growth. Thus, publications consider the specific features of innovative development and its impact upon labour productivity in small enterprises [5, 10]. Apparently, the investigated interrelations will be different within different economy sectors, which is proved by the researches [1, 17]. There are research papers dedicated to the role of innovations in agriculture [9], construction industry [11], electric-power industry [26], etc.

According to scientists' opinion and as confirmed by a number of research papers [2, 8, 12, 16], the inhomogeneity of the investigated effects also reveals itself depending on country's development level. At the same time, it should be noted that since the interpretation of the term "innovation" is ambiguous, scientists often use the development of information and communication technologies as the main indicator of innovative activity. These particular technologies serve as an innovative engine at the current stage of economic development in any country.

Thus, experts primarily focus on the interrelation of innovative activity and productivity at the level of enterprises, sectors and economies in general. Researchers pay much less attention to interregional differences, because this aspect is mostly relevant for big countries. For instance, the issue of overcoming regional barriers in innovative development is studied using China as an example [19].

Russia is notable for a great number of administrative and territorial units located within vast territory. The issues of studying the innovative factors of labour productivity growth in interregional context have been to a certain degree covered by Russian authors [7, 3 et al.]. However, it is necessary to perform further researches that would make it possible to approach the innovative regionalization in the country and the formation of regional innovative systems from the viewpoint of evaluating the internal cause-and-effect relations.

The goal of this research is to identify major factors, which have positive and negative impact on the regional level of labour productivity. For this purpose factors that characterize the innovative activity of organizations and the development level of small business in the investigated RF regions have been selected as independent factors. The specific feature of the research is the analysis of interregional differences of examined effects by the comparison of macro-regions.

2 Materials and Methods

The main research methods were regression and correlation analyses in order to determine how presumable changes (which can be managed at macro- and meso-levels of the parameters within the investigated system) influence the value of "output" characteristic, i.e. labour productivity.

At the first preliminary stage of econometric modeling there has been determined the following set of variables, which participate in the model of economic variables. As a performance indicator we have taken the indicator that describes labour productivity at federal subject level – GPR (gross regional product) per 1 employee. Following factors were attributed to independent variable factors:

X_1 – the level of organizations' innovative activity, which reflects percentage ratio of innovation-active organizations to their total number.
X_2 – the share of organizations, which performed technological innovations in %, calculated similarly to X_1, but according to the limited number of economic activities;
X_3 – the internal costs for scientific researches and developments, calculated per 1 employee, ths. Rubles;
X_4 – the ratio of costs for organizations' innovative activity to the amount of shipped goods (works, services), %;
X_5 – the share of innovative goods (works, services) within their total amount, %;
X_6 – the share of employees engaged in small business within their total amount, %;
X_7 – the ratio of small companies' turnover to organizations' turnover, %.

The first five variables present the information from statistical reports regarding the costs for innovative activity (X_3 and X_4) and the results consequently achieved (X_1, X_2 and X_5). The introduction of the last two variables is due to the fact that GRP is calculated taking into account the results of small business activity, although its share within statistical data up to 2021 was not indicated. At the same time, indicators that describe the innovative activity of organizations do not take the activity of small businesses into account.

For analysis purposes there were used data from 82 Russian regions for 2020, published on the website of Federal State Statistics Service (Rosstat) [23].

At the second a priori stage of modeling it has been determined that labour productivity in Russia has grown by 13.4% since 2014, which can be deemed as low growth rate (Table 1). The reasons for such a rate are connected with methodical specifics of determining this indicator (low annual average GDP growth rate, presented in USD in order to ensure data comparability).

Table 1. The dynamics of labour productivity in the RF

Indicator	2014	2015	2016	2017	2018	2019	2020	2020 in % to 2014
GDP in current prices, bln. Rubles	79030	83087	85616	91843	103862	109862	106967	135.3
GDB at purchasing power parity, bln. USD	3763.5	3526.2	3539.0	3807.1	4211.4	4283.5	4133.1	109.8
Number of employed, ths. People	71493	72425	72065	71843	71562	71064	69252	96.9
GDP at purchasing power parity, USD per 1 employee	52642	48688	49108	52992	58850	60277	59682	113.4

Compiled by the authors based on [21, 23].

The detected trends of slowdown in labour productivity growth are the consequence of many factors. Among them are not only such global events as Solow's productivity paradox [7], but also changes in operating conditions of Russian economy (sanctions and pandemic). According to the research, conducted by Federal Research Centre for Projects Evaluation and Consulting Services, significant impact is made by the reduction of organizations' activity level: their number within 2015–2019 decreased by 3% and the number of researchers decreased by 6.4% [24]. One of the reasons for that is the decrease of share of companies' internal costs for research and development works from

1.43 to 1.21% GRP. As a result, the share of innovative goods (works, services) within their total amount decreased from 8.44 to 5.27%.

At the third informational stage of the research work there was gathered necessary information with subsequent assessment of its homogeneity in order to exclude the objects with high degree of deviations (influential observations). The exclusion of these objects can significantly increase the predictive properties of the model. In particular, when considering GRP values per 1 employee in economy, which describes labour productivity, the following has been determined. The variation coefficient was 58.3%, which is a very high value. This situation is illustrated by Fig. 1 that provides data by Russian regions, grouped according to Federal Districts (FD).

The greatest difference is between the Sakhalin and the Tyumen regions – more than by 3.5 times. At the same time, their places among the ranking by innovative activity of 82 analyzed regions are 64 and 49 accordingly; and by internal costs for scientific researches and developments per 1 employee – 41 and 22 accordingly. The reason for such deviations is the high share of oil-and-gas field developments within the economy of these regions. The exclusion of these regions has reduced the variation coefficient to the acceptable level – 39.4%.

The Chechen Republic, Dagestan and Ingushetia (North Caucasian FD), which have an extremely low level of innovative activity, have been excluded, as well (Fig. 2). The Chukotka and the Jewish Autonomous Regions (Far Eastern FD) have been excluded because of the internal costs for scientific researches and developments per 1 employee, since data for these regions was not published.

At the fourth stage of the specification of the developed model we performed the regression analysis with the use of MS Excel "Regression" tool.

The analysis of regression statistics has shown that standard error after the exclusion of 8 federation subjects had reduced from 221 to 142 or by more than 1.5 times. Multiple correlation coefficient "R" was 0.623. This means that the relation between GRP per capita and all selected factors is average. R^2 was 0.39. In other words, change of GRP per capita is influenced by all considered factors together by 39%. This level can be deemed acceptable, because this research did not take into account the specific features of sectorial structure of regions' economy, composition of types and innovations on their territory. Apart from that, companies' innovative activity can also have negative impact upon redistribution of resources within the company, which is not monitored in course of statistical survey [6].

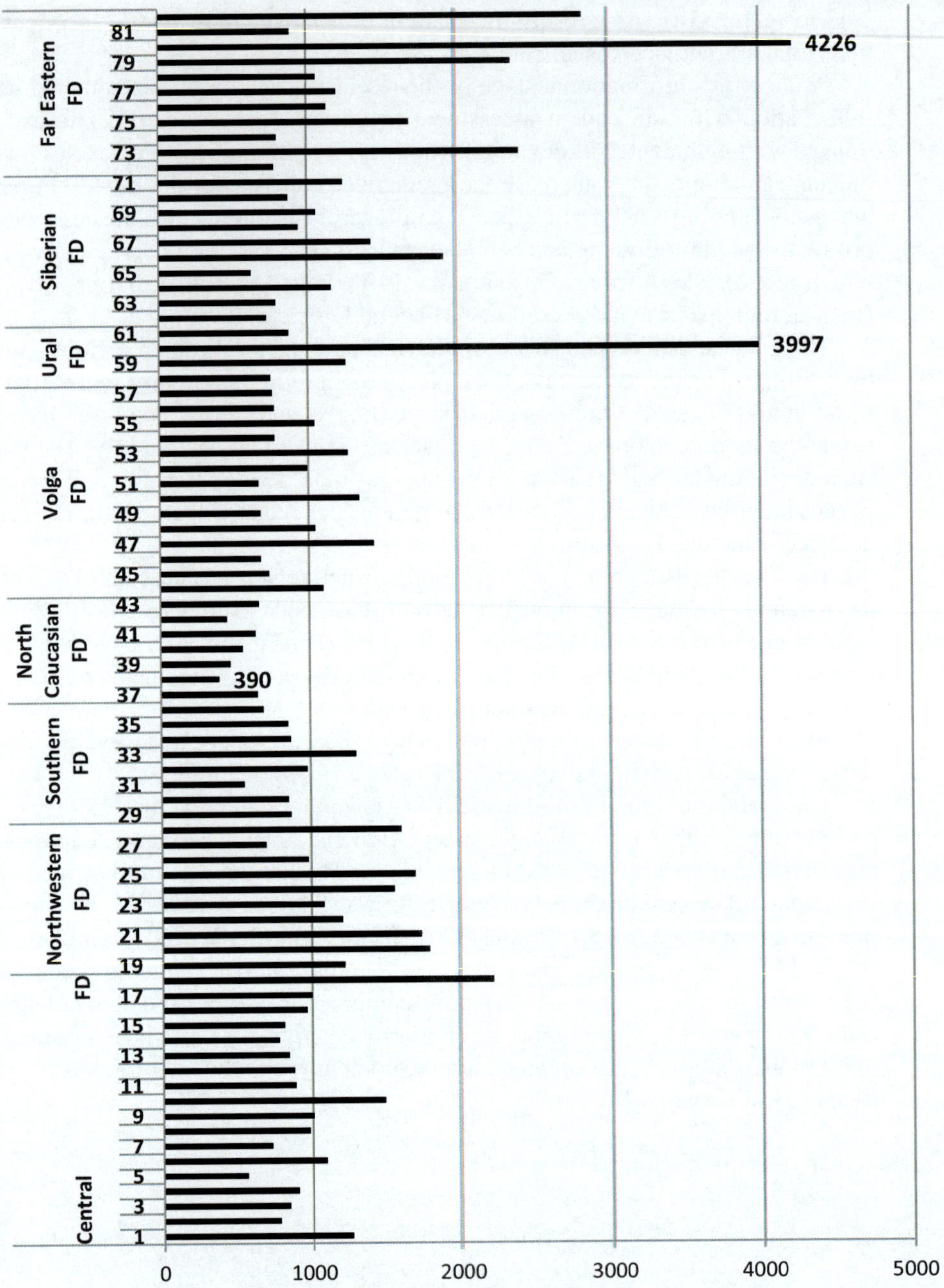

Fig. 1. GRP per 1 employee by the regions of the RF, ths. Rubles.

Next, the selected factors were checked for multicollinearity. For this purpose correlation analysis with the use of MS Excel "Correlation" tool was performed. The results of this analysis are presented in Table 2.

The availability of correlation coefficients by 0.8 modulus confirms multicollinearity of X_1 and X_2 variables. Since the performance indicator is much more influenced by X_2, it has been decided to exclude X_1 variable from further analysis (Fig. 2).

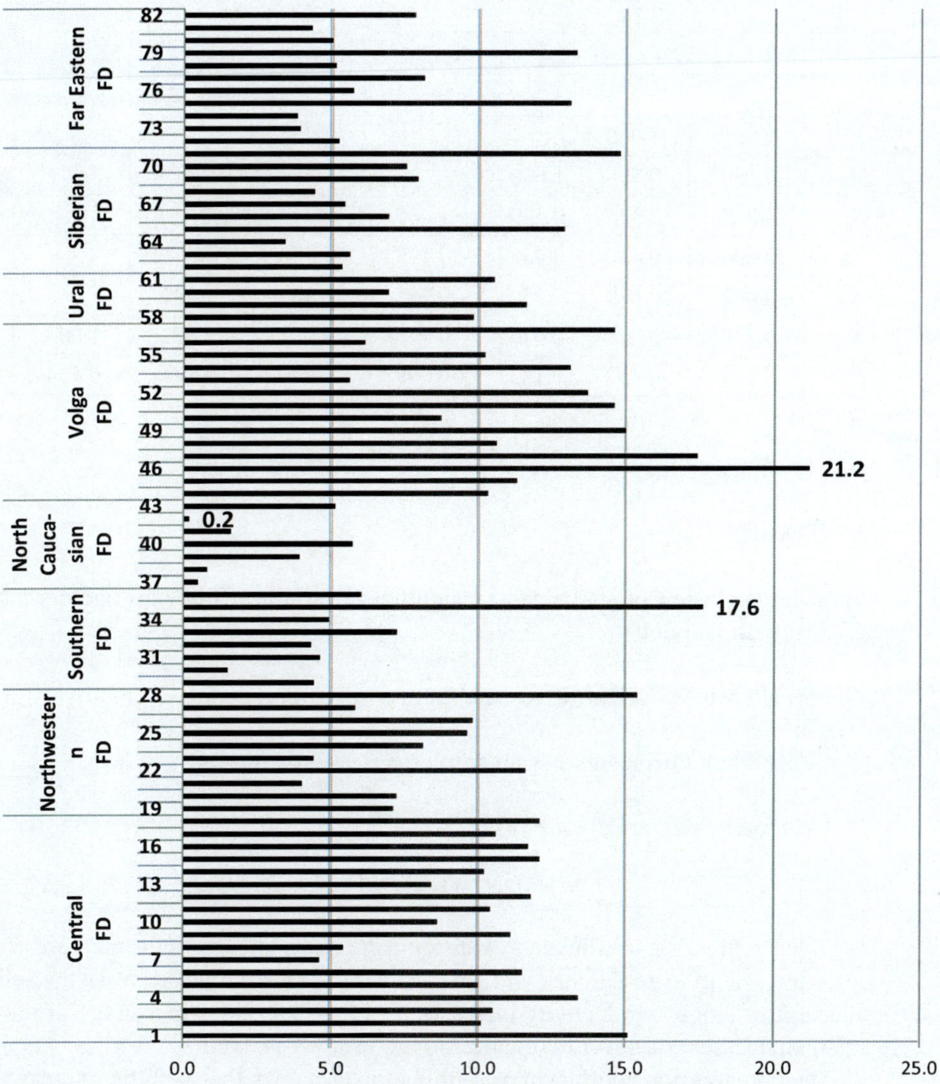

Fig. 2. The level of companies' innovative activity by the RF regions, %

Multiple correlation coefficients R and R^2 have not changed significantly in course of recalculation.

Having estimated the statistical significance of regression equation coefficients, we took the decision to exclude factors x_2, x_4, x_5, x_6 from the model, because their t-statistics is less than tabulated value. This is also confirmed by the analysis of p-values, which are more than 0.05 for the specified variables.

Table 2. The results of correlation analysis

	Y	X_1	X_2	X_3	X_4	X_5	X_6	X_7
Y	1.000							
X_1	0.061	1.000						
X_2	0.159	0.798	1.000					
X_3	0.380	0.375	0.539	1.000				
X_4	0.006	0.410	0.435	0.473	1.000			
X_5	−0.082	0.679	0.559	0.255	0.482	1.000		
X_6	0.330	−0.015	0.036	0.143	−0.081	−0.072	1.000	
X_7	−0.478	0.162	0.133	−0.099	0.000	0.174	−0.434	1.000

3 Results

After the exclusion of statistically insignificant variables following factors have been left within the model:

- internal costs for scientific researches and developments per 1 employee, ths rubles (x_3);
- ratio of small businesses' turnover to organizations' turnover, % (x_7).

Eventually, the model looks as follows:

$$Y = 1187.11 + 13.15x_3 - 6.03x_7$$

"Net" regression coefficient a_3 with x_3 factor shows that with the increase of internal costs for scientific researches and developments per 1 employee by 1 ths rubles, the indicator of labour productivity increases by 13.15 ths rubles in average, if the ratio of small businesses' turnover to organizations' turnover is fixed.

"Net" regression coefficient a_7 with x_7 factor shows that with the increase of ratio of small businesses' turnover to organizations' turnover by 1%, GRP per 1 employee will averagely decrease by 6.03 ths rubles, if internal costs for scientific researches and developments per 1 employee are fixed.

Changes of regression statistics with the reduction of number of independent variables are shown in Table 3.

Multiple R amounts to 0.58. This means that the relation between GRP per 1 employee, internal costs for scientific researches and developments per 1 employee and the ratio between small businesses' turnover to organizations' turnover is average.

Table 3. Change of regression statistics

Indicator	Indicator value		Indicator deviation
	With 6 factors	With 2 factors	
Multiple R	0.6228	0.5837	−0.0391
R^2	0.3879	0.3407	−0.0472
Normalized R^2	0.3331	0.3221	−0.0110
Standard error	341.36	344.16	2.80
Observations	74	74	X

R^2 value = 0.34 means that the change of labour productivity is determined by the dynamics of investigated factors in total by 34%.

The calculation of partial elasticity coefficients has shown the following. If internal costs for scientific researches and developments per 1 employee increase by 1%, labour productivity increases by 0.1% in average, if other model factors remain fixed. At the same time, the increase of ratio of small businesses' turnover to organizations' turnover by 1% will generally result in GRP per capita reduction by 0.221%.

In order to check the hypothesis regarding the existing regional diversity of the effects under investigation we have performed the regression analysis in the context of Russian macro-regions. Thus, Macro region I included Northwestern and Central FD, Macro region II – Southern and North Caucasian FD, Macro region III – Volga, Ural and Siberian FD. Regression statistics, obtained as a result of calculations, is provided in Table 4.

Table 4. Change of regression statistics by macro-regions

Indicator	Indicator value			
	Macro region I	Macro region II	Macro region III	The RF in general
Multiple R	0,75305	0,45726	0,61922	0,58370
R^2	0,56709	0,20909	0,38343	0,34070
Normalized R^2	0,53246	0,01136	0,33205	0,32213
Standard error	274,61	239,92	252,92	344,16
Observations	28	11	27	74
Coefficients: Y	1182,545	889,283	1342,949	1187,113
X_3	15,646	6,337	5,816	13,147
X_7	−4,662	−5,410	−8,392	−6,028

The obtained data confirm significant territorial differences within investigated interrelations. It should be noted that the Far East Federal District was excluded from calculations due to insufficient observations (8 subjects after the exclusion of the Sakhalin region, the Chukotka and the Jewish Autonomous Regions) and specific economic structure of this region.

4 Discussion

The obtained results of research have actually confirmed the conclusions of two investigations, carried out by the Chinese scientists: 1) the innovative activity of organizations leads to redistribution of their resources, which results in the reduction of economy-wide innovative effect [6], 2) in the context of macro-regions we can observe changes of innovative activity impact upon labour productivity, as compared to the data for the country as a whole [19].

As far as redistribution of resources is concerned, it can be noted that the implementation of advanced technologies definitely leads to the growth of labour productivity and the reduction of companies' employees. Unemployed personnel should seek for another job, which frequently results in changing not only occupation, but also profession. Low correlation coefficient between X_3 and X_7 (0.1327) shows that there is almost no interrelation between the growth of internal costs for scientific researches and developments per 1 employee and the growth of small companies' turnover in relation to organizations' turnover. Moreover, it should not be overlooked that the growth of GRP share, produced by small business, has negative impact upon changes of labour productivity within the economy of labour productivity in average. The reason for that are its weaker innovative activity [24] and the efficiency of utilization of engaged labour power.

As far territorial differences of the investigated relations as a consequence of inhomogeneous regional development are concerned, we can note the following. Within Macro region I there exists the closest connection between GRP per 1 employee, internal costs for scientific researches and developments per 1 employee and percentage ratio of small businesses' turnover to organizations' turnover. Changing these two independent variables by 56.7% explains GRP dynamics in Northwestern and Central FD. It is quite understandable, because of higher degree of development of high-technology and knowledge-intensive industries in these regions, as well as the concentration of innovative economic activity in Macro region I, and specifically in Moscow and St. Petersburg agglomerations.

Much worse is the situation in Macro region II, where there is almost no interrelation between the investigated parameters, although 3 republics (North Caucasian FD) were excluded from calculations due to their low innovative activity level. The difference from Macro region I is primarily in specific features of industrial research intensity. Obviously, one of the reasons for that is "resource-based" nature of economy in the subjects of Macro region II, with high share of such economic sectors as mining and agriculture (which can also be considered as resource-based economy sector). Their share is higher than the share of processing industries, where the possibilities for innovative development are wider.

Thus, within the subjects of Macro region II there should be formed a specific ecosystem of innovative development, which can be of supra-regional nature [20]. This ecosystem, comprising infrastructure, standards, regulatory norms is aimed at the development of territories' innovative potential and the promotion of specific regional innovative paths. By the creation of this ecosystem it is reasonable to apply the principles of "catching-up development" and polycentricity.

The principle of "catching-up" development is aimed at some compensation of the peculiarities of regional specialization by creating the drivers for innovations and economic development. This can be achieved with the help of different instruments, such as direct and indirect subsidies for encouraging the activity in the field of scientific researches, as well as different ways of initiating innovations in strategically valuable sectors.

The principle of polycentricity is aimed at the creation of innovative framework within the territory. The need for such framework is related to the existing spatial limitations in course of distributing innovations within the vast territories of Russian regions. The extension of network and strengthening the significance of scientific and innovation-technology centers with simultaneous increase of state supportive measures are able to become a basis for systemic implementation of innovative developments, technological diffusion within regional and industrial complexes.

5 Conclusion

Thus, keeping in mind the objective of the present research, which aims at the estimation of interrelation between labour productivity, innovative activity of organizations and the level of small business development in Russia in general and in the context of macro regions, we can state the following.

The interrelation of factors under investigation is of dual character. The implementation of advanced technologies leads to the growth of social labour productivity, but simultaneously results in redistribution of labour force into small business sector, the expansion of which is a decreasing factor for labour productivity at macro level.

The relation between indicators under analysis (GRP per 1 employee, internal costs for scientific researches and developments calculated per 1 employee, ratio of small businesses' turnover to organizations' turnover) should be estimated as average. Change of labour productivity is determined by the dynamics of independent factors under consideration by 34% in total.

The closeness of this interrelation within the selected macro regions changes significantly. The closest interrelation is in macro region, which comprises Northwestern and Central FD and is explained by the concentration of high-technology and knowledge-intensive industries, as well as the availability of scientific and financial centers.

The interrelation between the investigated parameters within the macro region that comprises FD of the south of Russia, is very weak. This can be primarily explained by a more distinct "resource-based" nature of the economy of these regions, as well as by lower industrial resource intensity. From the practical point of view, this situation requires creating new scientific and innovation-technology centers and increasing their significance, as well as developing new approaches to the formation of a specific innovative ecosystem of this territory, which is able to ensure "catching-up" development in the field of new technologies.

Ensuring sustainable development of Russia with GDP growth rates surpassing worldwide average growth rates can be achieved not only by the increase of internal costs for scientific researches and developments per 1 employee, but also by target-focused increase of these costs in small and medium-sized businesses, which manufacture import-substituting products and expand export potential of the country. At the same time, it is necessary to continue supporting small businesses, which ensures the employment of population, although not highly-effective.

This problem needs further studying with more fundamental analysis of factors, which have not been taken into consideration in the current research. Among them are: the specific features of sectoral structure of regional economies, composition of types and innovations on the territory of these regions.

References

1. Abramova, N., Grishchenko, N.: ICTs, labour productivity and employment: sustainability in industries in Russia. Procedia Manuf. **43**, 299–305 (2020)
2. Afonso, O., Pinho, M.: How to reverse a negative asymmetric labor productivity shock in the European Union? A directed technical change analysis with fiscal and monetary policies. Math. Soc. Sci. (2022, in press). Journal Pre-proof
3. Balatsky, E.V., Ekimova, N.A.: Internal sources of labour productivity growth in Russia. The World of New Economy **14**(2), 32–43 (2020)
4. Basovskaya, E.N., Basovskiy, L.E.: The problem of productivity factors in the economy of modern Russia. Nauchnye issledovaniya i razrabotki. Ekonomika **6**, 20–26 (2018)
5. Baumann, J., Kritikos, A.S.: The link between R&D, innovation and productivity: are micro firms different? Res. Policy **45**(6), 1263–1274 (2016)
6. Dai, X., Sun, Z.: Does firm innovation improve aggregate industry productivity? Evidence from Chinese manufacturing firms. Struct. Chang. Econ. Dyn. **56**, 1–9 (2021)
7. Dementiev, V.E.: Productivity paradox in regional dimension. Eco. Reg. **15**(1), 43–56 (2019)
8. Dieppe, A., Francis, N., Kindberg-Hanlon, G.: Technological and non-technological drivers of productivity dynamics in developed and emerging market economies. J. Econ. Dyn. Control **131**, 104216 (2021)
9. Djoumessi, Y.F.: What innovations impact agricultural productivity in Sub-Saharan Africa? J. Agric. Food Res. **6**, 100228 (2021)
10. Edeh, J.N., Acedo, F.J.: External supports, innovation efforts and productivity: estimation of a CDM model for small firms in developing countries. Technol. Forecast. Soc. Chang. **173**, 121189 (2021)
11. Greco, M., Grimaldi, M., Locatelli, G., Serafini, M.: How does open innovation enhance productivity? An exploration in the construction ecosystem. Technol. Forecast. Soc. Chang. **168**, 120740 (2021)
12. Hamrouni, D.: International diffusion of knowledge labor productivity and catching up between North and South. Int. Rev. Econ. Financ. **77**, 170–178 (2022)
13. Khanna, R., Sharma, C.: Testing the effect of investments in IT and R&D on labour productivity: new method and evidence for Indian firms. Econ. Lett. **173**, 30–34 (2018)
14. Kurt, S., Kurt, Ü.: Innovation and labour productivity in BRICS countries: panel causality and Co-integration. Procedia. Soc. Behav. Sci. **195**, 1295–1302 (2015)
15. Labor productivity: new trends, old problems. https://www.finam.ru/analysis/forecasts/proizvoditelnost-truda-novye-tendencii-starye-problemy-20210831-163836/ (2021). Accessed 01 Feb 2022

16. (Maggie) Fu, X., Bao, Q., Xie, H., Fu, X.: Diffusion of industrial robotics and inclusive growth: Labour market evidence from cross country data. J. Bus. Res. **122**, 670–684 (2021)
17. Moussir, C.-E., Chatri, A.: Structural change and labour productivity growth in Morocco. Struct. Chang. Econ. Dyn. **53**, 353–358 (2020)
18. Nikonova, Y.I.: Study of the relationship of innovation and economic growth of national economies. Scientific-Methodological E-Journal Concept **15**, 2001–2005 (2016)
19. Pan, W., Xie, T., Wang, Z., Ma, L.: Digital economy: An innovation driver for total factor productivity. J. Bus. Res. **139**, 303–311 (2022)
20. Preobrazhensky, Y.V.: Macro-regions as supraregional innovative systems: the problem of the flow of knowledge. News of Saratov University. New Series. Series: Earth Sciences **20**(1), 17–22 (2020)
21. Ranking of countries by GDP in 2020. https://svspb.net/danmark/vvp-stran.php. Accessed 01 Feb 2022
22. Ranking of countries in the world by standard of living. https://basetop.ru/rejting-stran-mira-po-urovnyu-zhizni-2021/ (2021). Accessed 01 Feb 2022
23. Regions of Russia. Socio-economic indicators. https://rosstat.gov.ru/storage/mediabank/Region_Pokaz_2021.pdf (2021). Accessed 01 Feb 2022
24. Scientific Research Institute – Federal Research Centre for Projects Evaluation and Consulting Services. Innovative development of the Russian Federation in 2019. https://www.miiris.ru/digest/analitika_RF.pdf (2022). Accessed 01 Feb 2022
25. Semenova, N.N.: Assessment of the innovations impact on the pace of economic development and labour productivity in the Republic of Mordovia. Izvestiya Vysshikh Uchebnykh Zavedenii. Povolzhskiy region. Social Sciences **2**(54), 126–137 (2020)
26. Vu, K., Hartley, K.: Effects of digital transformation on electricity sector growth and productivity: a study of thirteen industrialized economies. Util. Policy **74**, 101326 (2022)

Flour Export as a Factor of Sustainable Development of Rural Areas: Problems, Prospects

Natalya Vorobyeva(✉), Daria Gracheva, Svetlana Vaytsekhovskaya, Yuliya Orel, and Elena Pupynina

Stavropol State Agrarian University, Stavropol, Russia
vorobeva1979@mail.ru

Abstract. The question of sustainable development of rural areas is needed in analysis of production capabilities of wheat or rye flour and orientation of producers to the foreign market. At that point, the study of the problems and main trends of the flour-grinding segment in the global space becomes actual. The research included an analysis of wheat and meslin export production, also products of their processing in the global market. A detailed study was carried out on the statistical data of the Russian Federation. As an example, we used the southern region of the country - the Stavropol Territory, which specializes in grain crops cultivation and its processing. The purpose of the scientific study is the assessment of opportunities for rural areas development. The study is based on the identification of factors that have a positive and negative impact on the export potential of flour products. It was founded that nowadays the Russian Federation occupies the 10th position among global suppliers of wheat and rye flour. The Stavropol Territory is one of the three leaders in the export of flour products and has a significant potential to increase. Despite the fact that Russia by itself concedes in export volume to Turkey, Kazakhstan, and Uzbekistan, the positions of the Stavropol Territory in this segment are being strengthened. To maintain the positive trend, it is necessary to expand the exports geography in perspective areas (the countries of the Middle East, Africa, Southeast Asia), to improve the exports support and to harmonize tariff and non-tariff measures between countries. The results of the correlation-regression analysis showed that the price does not have a significant impact on the flour-grinding products export. Therefore, there is a background to increase the export deliveries with government support regardless of price and sustainable status of rural areas. The novelty of the results lies in assessing the influence of main factors on the volume of flour exports in the Stavropol Territory and characterizing the complex of problems existing in this context. New directions for the agricultural products supply to the foreign market are substantiated.

Keywords: Wheat flour · Rye flour · Wheat · The Stavropol Territory · The Russian Federation · Export · Price · Cereals · Global market · Production

I. Samoylenko (Ed.): AMBT 2022, LNNS 582, pp. 124–131, 2023.
https://doi.org/10.1007/978-3-031-20803-4_13

1 Introduction

The development of international trade has a certain impact on the country specialization in the global market and has transformed the structure of producers, exporters and consumers of grain products. Wheat, as the main grain crop, is considered as a necessary element in the diet of many nationalities. Many Russian and foreign scientists consider that the world demand on wheat flour is limited, because many countries prefer to import grain instead of flour. And the import duties are usually very high, volumes are quota, as a result the entry into new markets for Russian exporters may be difficult [8, 9, 12]. Despite the decline of global consumption of flour products (which is observed during the pandemic since 2020), the dynamics of exports is in a limited range and there is no significant reduction in supplies. Most countries are still exporting flour, trying to support local producers and create favorable conditions for the sustainable development of rural areas.

From the one side, there is the grain segment importance in the world space, for the other, there is the export potential importance for the country as a component of the sustainable development of rural areas. The state of the global market of grain crops, including wheat, the analysis of geography, exports and imports dynamics are detailed in international studies [2, 3, 15, 16, 19]. These articles are focused on the competitiveness factors, the import volumes dependence on the price [2–4, 10, 11], the specific features of each country, the definition of effects by increasing production [6, 8, 17], the protectionism practice [1, 18], certification and standardization questions [7]. Great emphasis is placed on the analysis of comparative advantages in agriculture of different countries [2, 3, 13]. Many scientists substantiate the conclusions about the volatility of trends in the export of grain [2, 3, 13]. The studies above became a sufficient basis for the next researching in this area. However, there are still unresearched issues, such as the development of the world market of grain processing products, the wheat flour in particular, development factors and problematic aspects associated with the intensification of the export potential of flour products as an indicator of the development of rural areas. These factors determined the relevance of research, its theoretical and practice implementation. In this regard, the purpose of the study is to identify trends of the development of wheat or rye flour exports, taking into account the obstacle factors for export of grain processing products and for the sustainable development of rural areas.

2 State and Development of Grain Products and Their Processed Products in the Global Market

Under the influence of natural and climatic factors and the socio-cultural preferences of people in different parts of the world, a certain specialization of the crop production has developed.

Currently, the main volume of produced wheat is concentrated in the countries of the European Union (137.5 million tons), China (136 million tons) and India (108 million tons). Russia, according to quantitative estimates, remains in the fourth place (73.5 million tons, 10.4%), ahead of the United States, which closes the top five countries.

However, specialization in certain types of grain is not always translated into export competitive advantage in the world market. Thus, China and India, the world leaders in the gross harvest of wheat, almost completely use the produced volumes of cereal for domestic consumption and are forced to import grain. The leading exporters in 2020 were Russia (37.3 million tons), USA (26.2 million tons), Canada (26.1 million tons), France (19.8 million tons) and Ukraine (18.1 million tons). These countries supplied 127.5 million tons of wheat to the world market this year, or total two thirds of the world exports volume.

The geography of supplies of wheat or rye flour has a slightly different structure: there is a discrepancy between the roles of countries wheat-exporters and countries supplying flour products. Thus, Turkey (2990.4 thousand tons), Kazakhstan (1742.6 thousand tons), Germany (900 thousand tons), Uzbekistan (725.1 thousand tons) and Argentina (598.4 thousand tons) are the leaders in the supply of these products for export. Russia occupies only the tenth position in the flour export. The main reason for the country's low rating is the minimum world prices for wheat or rye flour during 10 years. Thus, the most competitive are those market participants that are able to generate high profitability due to large volumes of trade, high quality products, or due to government support.

World prices for wheat before 2015 were an rather higher than during the more productive periods. The lowest price level was fixed in 2016, when the supply of crops on the world market increased significantly compared to the previous periods (Table 1). The highest level of wheat prices is observed in 2018, which is associated with a supply and consumption level decrease in the world wheat market. The weakening of the ruble also strengthened the position of Russian exporters in foreign markets, when Russia ousted France out of the large Egyptian market. However, the price per ton for wheat or rye flour in the leading countries in 2020 was: in Turkey - $318, Kazakhstan - $281, Germany - $384, Uzbekistan - $301, Argentina - $325, Ukraine - $269. According to the average prices of the world market, the cost of wheat or rye flour in Russia ($312) is considered uncompetitive, exceeding prices of many countries. Secondly, the logistics chains of competing countries are more developed, and the transport costs will be lower due to the territorial proximity to the countries of East and Southeast Asia, the Middle East (for example, in Turkey).

Table 1. An average world export prices for agricultural products, USD/ton [5, 14]

Indicators	2015	2016	2017	2018	2019	2020
Wheat and rye flour export price in Russia	320.0	300.0	318.0	270.0	327.0	312.0
Wheat and meslin export price in Russia	186.0	166.0	175.0	192.0	201.0	212.0
Wheat and rye flour world price	405.0	370.3	321.4	372.7	393.0	415.0
Wheat and meslin world price	197.0	149.9	203.0	218.0	196.0	220.0

3 Development of the Export Potential of Rural Areas in the Russian Federation (on the Example of the Stavropol Territory)

Each Russian region has its own specific structure of agricultural products and export opportunities. This structure is based on natural and climatic conditions, proximity to sales markets, demand of produced raw materials (both in the region and beyond), transportation costs, production capacities, qualifications of personnel, the structure of crops. Together, all factors are forming the competitive advantage of the region and its production, contributing to the sustainable development of rural areas.

Given to the considerable length of the Russian Federation, many regions in it have a common border with foreign countries or are located nearby. For example, the Far Eastern Federal District (Primorsky Territory), the Siberian Federal District (Novosibirsk Region, Altai Territory, Kemerovo Region) are close to China. Due to the existing level of specialization, grain processing products can be the export basis to China in these regions.

In accordance to the |Russian Strategy of the Sustainable Development of Rural Areas and the federal project "The Export of Products of the Agro-Industrial Complex (AIC)", the question of stable socio-economic development of rural settlements and exports increase due to the high added value products can be resolved. Thus, on the one hand, the problem of the development of rural areas will be solved because of the development of agricultural producers, the growth of positive cash flow, revenue from products sales, profits, and on the other hand, the implementation of the President's message to increase the export of non-commodity products of the agro-industrial complex.

The export of Russian flour is characterized by unstable dynamics. This is evidenced by the fall in the value of exports in 2020 to $77.3 million, or 25% down. The drop in flour-grinding products exports is explained in reduction of China purchases. In addition to PRC, Russia's partner countries in flour imports are Georgia, Afghanistan and Iraq.

The Stavropol Territory is considered to be a highly developed agricultural region, with the largest agro-industrial complex in south of Russia. By implementing the programs above, the region can ensure the development of rural areas due to the effect of the grain processing products export. Therefore, the export of flour should become a strategically important export direction and, as a result, an increase of grain processing products, the purchase of new varieties of grain crops, an increase in payments to budgets of all levels, an increase in wages in rural areas.

Conducted studies have shown that the share of the Stavropol Territory in Russian exports of wheat or rye flour ranges from 7–10% (27–29 thousand tons), which indicates a large contribution of the region to the country's economy. In 2020, during the pandemic, the region's rating rose by 2 points and fixed at the 3d position. The share of the Stavropol Territory in wheat exports in the country in 2020 was 1.5% (Table 2).

Using the trend analysis of the wheat or rye flour export in Russia and in the Stavropol Territory, the low determination coefficients were obtained - 0.379 and 0.221, respectively. The results of the study indicate that the trends of export volumes growth or decline are non-predictable and forecasting becomes inappropriate.

A paired correlation-regression analysis between the volume of flour exports in the Stavropol Territory and the average export price in Russia showed that there is no close relationship between the factors, and price changes do not significantly affect the amount of exported flour, since the correlation coefficient was minus 0.519.

Table 2. Export indicators of wheat and meslin, and products of their processing in the Stavropol Territory [5, 14]

Indicators	2015	2016	2017	2018	2019	2020
Wheat and rye flour export in the Stavropol Territory, million dollars	9.9	5.5	5.0	6.5	8.4	7.5
Wheat and meslin export in the Stavropol Territory, million dollars	85.8	68.8	64.4	69.1	103.0	77.3
Wheat and rye flour export ranks of the Stavropol Territory	4	6	7	5	5	3
Wheat and meslin export ranks of the Stavropol Territory	4	6	7	7	6	7
Wheat and rye flour export in the Stavropol Territory, %	11.5	8.0	7.8	9.4	8.2	9.7
Wheat and meslin export in the Stavropol Territory, %	4.3	2.7	2.1	0.9	2.0	1.5

Therefore, the region should pay attention to other internal and external factors in the development of the Russian and the world flour markets. The advantage of the grain and flour milling segment of the Stavropol Territory is in its favorable economic and geographical position for neighboring importing countries (Azerbaijan, Armenia, Georgia, Turkmenistan, Kazakhstan). Among the weaknesses, we have identified: underdevelopment of innovative production and market infrastructure; technical barriers preventing the formation of flour consignment; low level of technical equipment of mills; lack of stable export channels; difficulty in obtaining timely and reliable information. The opportunities include access to new markets (Afghanistan, Iraq); increasing exports volume of environmentally friendly products from grain using organic farming technology; the increase in population in the countries of East and Southeast Asia, the Middle East, Latin America. The planned duties on wheat (export - up to 50 euros per ton, prohibitive - 50% of the customs value, but not less than 100 euros per ton) will help to stabilize flour prices and regulate wheat exports. And it may be possible to redistribute the structure between grain and flour. in favor to the last one. As a result of the tariff measures introduction, Russia will be able to bypass Turkey as an intermediary for wheat processing, and the Stavropol Territory will begin to supply wheat flour directly to Syria, Libya, Iran, as well as other countries in Asia and Africa. A good export opportunity for the Stavropol Territory is the capacious Egyptian market, to which Turkey currently supplies up to 3575 tons of flour from processed Russian grain. Threats can be: unfavorable world conjuncture; low agro-climatic potential, which has a negative impact on the exports development; shortage of wagons and high cost of wagon rent; formation of export monopolies; high transaction costs and transport tariffs; lack of transportation conditions unification for export.

The Stavropol Territory has a large production capacity (up to 1 million tons per year) for grain crops proceeding. That is, with a load of 300 thousand tons in 2020, the region has a reserve for wheat processing. In addition, the Government introduces mechanisms for permanent regulation of grain exports, which consists of the introduction of a "floating" duty on the export of raw materials abroad with the subsequent return of these funds as investments in agriculture. The introduced export duty will be determined depending on current export prices, but the final amount of the duty will be determined by the Ministry of Agriculture based on the contract on the stock exchange. Comparing the prices of flour and grain, it should be noted that in the future, the increase in the world market of Russian wheat flour is a way not only to ensure foreign exchange inflow to the industry, but also an incentive to further yields increase due to intensive and innovative factors. In modern conditions, the key to expanding exports is the foreign economic cooperation, so the promotion of wheat or rye flour should be accompanied by administrative and marketing support from the state. An example of such support is shown by countries that occupy a leading position in the flour export.

In Turkey, foreign trade operations with grain and processed products are carried out by companies authorized by the Turkish Grain Committee. This is a state structure, with very extensive power: from purchase prices regulation to compensation payments setting for exporters.

In Kazakhstan, the state also actively promotes the flour exports development, concludes interstate agreements and develops state support measures. In particular, a number of laws have been adopted (affordable loans, crop insurance, subsidies for plant protection) aimed at increasing grain production.

On the part of the Russian state authorities, support for grain processing enterprises can be carried out in the following areas:

- an assistance, including financial instruments, for participation in foreign exhibition and congress events; promotion of products through trade missions in the Russian Federation in foreign countries, the office of agriculture attache;
- discussing issues of product access to foreign markets in the course of bilateral negotiations; taking into account the interests of domestic producers in negotiations within the framework of free trade agreements between the members of the Eurasian Economic Union and third countries.
- involvement of interested companies in order to promote products to participate in intergovernmental commissions and working groups in trade cooperation.

The implementation of these measures will facilitate the access of Russian processing enterprises to export activities, simplify customs and administrative procedures and significantly increase the growth rate of flour exports. Also new agricultural producers will participate in export deliveries, and that will help to improve the financial results, to ensure profitable part of local budgets, to the growth of social spending and to the sustainable development of rural areas.

4 Discussion and Conclusion

The obtained studies show that in order to realize export opportunities, the country with its regions, on the one hand, must have huge reserves of wheat and production capacities for processing raw materials into flour. And, on the other hand, agricultural producers are trying to find a way out in pricing, competing with the world market prices, resisting the established quotas for the flour import of and other obstacles of the trade relations development. In conclusion of the study, it should be noted that in the future, the existing partnerships of the Stavropol Territory with foreign countries will not only remain, but there is a possibility to increase supplies to Egypt, Georgia, Abkhazia, Iraq and Afghanistan, Syria, Angola. China is considered to be a promising market with growing consumption of flour made in Russia. The region is characterized by positive factors (economic and geographical position, the availability of a raw material base, etc.) and problems (backwardness of innovative production and market infrastructure; technical barriers; low level of technical equipment of mills; difficulty in obtaining timely and reliable information). Nevertheless, the obtained results indicate the presence of the export potential of the Stavropol Territory, which will contribute to improve the population welfare and the sustainable development of rural areas.

Thus, by implementing the Strategy for the Sustainable Development of Rural Territories and the Federal Project "Export of Agro-Industrial Complex Products" the region, in our opinion, can ensure the development of rural settlements due to the effect of the grain processing products export. Therefore, the export of flour should become a strategically important direction and, as a result, an increase in the production of grain processing products, the purchase of new varieties of grain crops, an increase in payments to budgets of all levels, an increase in wages in rural areas.

The research materials can be required in case of substantiating the strategy of flour milling development as part of the sustainable development of rural areas. This study can be useful for agricultural producers in production programs developing which are aimed at expanding sales and strengthening export potential.

References

1. Barichello, R.: The COVID-19 pandemic: Anticipating its effects on Canada's agricultural trade. Canadian Journal of Agricultural Economics/Revue canadienne d'agroeconomie **68**(2), 219–224 (2020). https://doi.org/10.1111/cjag.12244
2. Benesova, I., Maitah, M., Smutka, L., Tomsik, K., Ishchukova, N.: Perspectives of the Russian agricultural exports in terms of comparative advantage. Agric. Econ. **63**(7), 318–330 (2017)
3. Benesova, I., Smutka, L., Hinke, J., Laputkova, A.: Competitiveness of mutual agrarian foreign trade of the post-soviet Countries. Ekonomie a Management **23**(3), 49–66 (2020)
4. Camanzi, L., Polino, M., Verza, M.: Market structure and supply chain strategies in the global agricultural commodity industry: a comparison between EU and USA. J. Glob. Bus. Adv. **13**(2), 249–269 (2020)
5. Export from Russia [e-re-source]. https://ru-stat.com/date-M201701-201712/RU/export/world/01
6. Generalov, I., Suslov, S., Bazhenov, R., Sibiryaev, A., Firsova, E.: Management system of grain production cluster of the region. In: IOP Conf. Series: Earth and Environmental Science. IOP Publishing, Atlanta, GA/US (2020)

7. Hlupak, Z.I.: Standardization and certification of grain in Ukraine and the European union countries. Sci. Horiz. **7**(80), 63–69 (2019)
8. Hossain, S., Delin, H.: Measuring economic impact in Korea, Japan, India, China, and ASEAN considering agricultural sectors: a dynamic CGE approach based on GAMS. Rev. World Econ. 10 (2021)
9. Krylatykh, E., Belova, T.: Export of Russian grain in the context of the formation of regional economic policy. Econ. Reg. **14**(3), 778–790 (2018)
10. Kusakina, O.N., Kazarova, A.Y., Kosinova, E.A., Agalarova, E.G., Aydinova, A.T.: Current state and development trends of the foodnet market in Russia. In: IOP Conf. Series: Earth and Environmental Science. IOP Publishing Ltd., Russia (2021)
11. Luo, P., Tanaka, T.: Food import dependency and national food security: a price transmission analysis for the wheat sector. Foods **10**(8), 1715 (2021)
12. Melnikov, A.B., Michailushkin, P.V., Popok, L.E.: Prospects for the development of advanced grain processing in Russia. Int. J. Econ. Bus. Adm. **8**(2), 276–282 (2020)
13. Mizik, T., Gál, P., Török, Á.: Does agricultural trade competitiveness matter? the case of the CIS countries. Agris On-line Pap. Econ. Inform. **12**(1), 61–72 (2020)
14. TRADE MAP [e-re-source]. TRADE MAP. https://www.trademap.org/Index.aspx
15. Tyupakov, E.K., Reznichenko, D.S., Klochko, E.N., Vertiy, M.V., N.V.: Development of the Grain market export potential. Int. J. Resent Technol. Eng. **8**(3), 7146–7152
16. Tyupakov, K.E., Reznichenko, S.M., Shichiyakh, R.A.: Priority directions for regional grain market development. Int. J. Eng. Adv. Technol. (IJEAT) **1**(9), 3385–3387 (2019)
17. Vorobyova, N.V., Gracheva, D.O., Bannikova, N.V., Telnova, N.N., Cheremnich, M.B.: Perspective directions of the agri-food market development in the system of bilateral foreign trade relations between Azerbaijan and the Stavropol Territory. In: IOP Conf. Series: Earth and Environmental Science. IOP Publishing Ltd., Russia (2021)
18. Xie, C., Grant, J.H., Boys, K.A.: Do state-owned enterprises benefit more from China's cereal grain tariff-rate quota regime? Can. J. Agric. Econ. **67**(4), 397–416 (2019)
19. Yesakhmetova, L., Dzholdasbaeva, G., Sauranova, M.: Food security peculiarities: factors affecting export capacities of one of the major word grain producers. J. Secur. Sustain. Issues **10**(1), 321–328 (2020)

The Development of Polycentric Agglomeration and the Non-agglomeration Territory in the Economic Space of a Region

Elena Koroleva(✉) and Natalia Zelepukina

Samara State University of Economics, 141, Soviet Army Str., 443090 Samara, Russia
korol388@mail.ru

Abstract. The uneven spatial development, features and problems of urbanization are important topics of modern research in the field of sustainable development of territories. The authors of the article carried out an empirical verification of the hypothesis, which suggests that within the same administrative region, the agglomeration territory is characterized by better dynamics of the economic space compared to the peripheral territory. The objects of study were the third largest in Russia Samara-Togliatti agglomeration with a polycentric spatial structure and the territory outside the agglomeration (far periphery) of the Samara region. The author's method of comparative analysis included, among others, the stage of selecting groups of indicators that characterize the economic space of the compared territories and, at the same time, are provided with official Russian statistics. The results of the study were quite unexpected: they showed that for the agglomeration, the dynamics in the group of indicators "saturation of the economic space with the activities of economic agents" is worse compared to the periphery. The reasons for this, according to the authors, could be the "deagglomeration" effect, as well as the phenomenon of "immunity of peripheral territories", which prevents the decrease in their viability and resilience.

Keywords: Spatial development · Region · Polycentric agglomeration · Non-agglomeration territory · Periphery · Comparative analysis · Russia

1 Introduction

Russian regions are undergoing significant transformations in spatial development. In the studies of the economic space of the region, there are the following major theoretical developments: the theory of central places, the theory of development poles and growth centers, the concept of a supporting framework, a sub-regional approach [1].

Differences in scientific approaches to the study of economic space cause the presence of several points of view on its most important characteristics – properties and structure. Among the fundamental features of the Russian economic space, Minakir & Demyanenko [2] highlights the fragmentation of the economic space and polarized centralization, emphasizing the special role of urban agglomerations in the formation and organization of economic areas closing on them. Agglomerations condense the economic

I. Samoylenko (Ed.): AMBT 2022, LNNS 582, pp. 132–139, 2023.
https://doi.org/10.1007/978-3-031-20803-4_14

space, create spatial clusters of economic activity, transmit innovations, resources and institutions to the space surrounding megacities.

Agglomeration centripetal trends are determined by enormous opportunities for entrepreneurs, the population, and authorities. These opportunities arise due to the agglomeration effect [3, 4]. This fact has been confirmed by a vast empirical material: for different countries, estimates of an increase in the labor productivity as a result of the doubling of the city, according to Kolomak [5], range from 3 to 16%.

In studies on spatial structure, as a rule, two main types (models) are distinguished: (a) a center-peripheral (concentric) type based on a clear separation of the dominant economic center and the peripheral zone as it moves away from it; (b) a polycentric type, where there are several centers, usually urban ones, between which a system of relations is formed. As for the center-peripheral type, we point out that the center, as a rule, is a large city, the nearest area of influence of which becomes an agglomeration. As for the peripheral territory, Russian researchers have recently been paying increased attention to their studies focusing on the complex problems of the Russian periphery. From the standpoint of the economic and geographical approach, Tsarev gives two definitions of it that are different in their meaning: 1) the territory defining the outer edge of the influence area of the center and adjacent to it from the inside (periphery - the edge of the center); 2) the totality of external territories located outside the influence area of the center, adjacent to it from the outside (periphery - anti-center) [6]. Kazakov, in his works devoted to the methodology of the system-diagnostic approach to the study on issues peripheral territories, systematizes and reveals their key properties and, on the basis of remoteness from the center, distinguishes the far, middle and near periphery [7].

Turning to the second (polycentric type) of spatial structure, we point out the existence of several approaches to the definition of the concept of "polycentricity", for instance, from the point of view of the morphological structure of the region or from the position of the functional approach. In European studies and practice of regional policy, the widely used concept of polycentrism in modern conditions receives a new interpretation and new accents [8]. The results obtained by Wessel, Ouwehand, van Oort & Cortinovis indicate a greater role of urbanization compared to polycentricity and dispersion than it was previously assumed, which allows us to consider it as a less confident support and panacea for reducing regional economic inequality [9].

Based on the above, it can be stated that in the study on the economic space of a region (subject of the Russian Federation) within the administrative boundaries, it is advisable to take into account the following main characteristics of its structure and properties: (1) the presence of two types of territories - agglomeration and non-agglomeration territories; (2) positioning of the agglomeration area as the center, hence the non-agglomeration area as the outer (far) periphery; (3) the complex spatial structure of the territory of each type in the coordinates "monocentricity-polycentricity" and "concentration (localization)-dispersion" which implies several combinations of two spatial dimensions.

The relevance of the research field is determined by the focus on agglomerations as growth poles in the Spatial Development Strategy of Russia [10]. Of particular interest in this case is the choice of an agglomeration area with a polycentric structure as a research object.

2 Materials and Methods

The authors' hypothesis is: do agglomeration territories with a polycentric structure have better characteristics of the dynamics of economic space compared to non-agglomeration (peripheral) territories not taking into account the characteristics of their structure?

The purpose of the study is to verify this hypothesis on the basis of empirical results obtained using the authors' methodology of the comparative analysis of the development of agglomeration of regional and non-agglomeration (peripheral) territories in the economic space of the region (subject of the Russian Federation). Research objectives are: (1) selection and characterization of objects of comparative analysis - agglomeration and non-agglomeration territories within the administrative boundaries of a specific Russian region (subject of the Russian Federation); (2) development of the authors' methodology for the comparative analysis of the development of agglomeration and non-agglomeration territories in the economic space of the region; (3) implementation of the authors' methodology and discussion of the results obtained in relation to the proposed hypothesis.

Two types of territories were selected as objects of the comparative analysis - agglomeration and non-agglomeration territories of a specific subject of the Russian Federation - the Samara region. The agglomeration area – the Samara-Togliatti agglomeration (STA) is the third largest urban agglomeration in Russia (after Moscow and St. Petersburg). The most important feature of the STA is polycentricity, this property is unique for the agglomerations of Russia; the formed cores of the agglomeration are two cities (Samara and Togliatti), the core at the stage of formation is the town of Syzran. The composition and structure of the STA are not officially approved in the regulations, so we will adhere to the composition of the STA according to the Strategy of Socio-Economic Development of the Samara region for the period up to 2030 in relation to the grid of the administrative-territorial division of this subject of the Russian Federation [11].

At the moment, the STA includes the following municipalities: 1 urban district with an inner-city division (Samara), 9 inner-city districts, 7 urban districts (Togliatti, Syzran, Novokuibyshevsk, Chapaevsk, Zhigulevsk, Oktyabrsk, Kinel), 9 municipal districts, 9 urban settlements, 112 rural settlements. The total population of the STA on 01.01.2020 is estimated at 2744306 people. The non-agglomeration area consists of two urban districts (Pokhvistnevo and Otradny) and 18 rural municipal districts respectively. The non-agglomeration area can be positioned as the far periphery.

The authors' methodology of the comparative analysis involves several stages:

1. Selection of indicators used to characterize the spatial development of the compared territories (Table 1).

Table 1. Indicators used to characterize the spatial development of the compared territories

Composition of indicators by groups			
Saturation of the economic space with activities of economic agents	Development of the economic space	Connectivity of economic space	Conditions of social development of territories
• goods of own production shipped, works and services performed by their own forces per capita (without small business entities); • investments in fixed assets per capita	• population density; • migration balance per 1000 population	• length of roads per 1 km^2 of territory	• the average monthly nominal accrued salary of employees of organizations

2. Collection and verification of initial data for each municipality of the Samara region using information from the official website https://www.gks.ru.
3. Calculation of values of these indicators per capita living on the territory of the relevant municipality or on 1 km^2 of its area.
4. Calculation of the average values of each indicator for the STA and for the non-agglomeration area.
5. Comparison of the obtained average values for STA and for non-agglomeration territories in dynamics for the selected study period.

All calculations are performed using the Microsoft Office 2019 Professional Plus (Excel) application software package.

3 Results

The comparative analysis was implemented for the selected period 2016–2019. The analysis of the chain growth rate of investment volumes for two types of territories (Fig. 1) for the period from 2016–2019 demonstrated that initially the growth rate of investment in the STA was higher. The gap between the two types of territories was 12%.

However, since 2017, we have seen a sharp jump in the rate of investment growth in peripheral municipalities (the gap was 209%). Following, in 2018 and 2019, this trend continues.

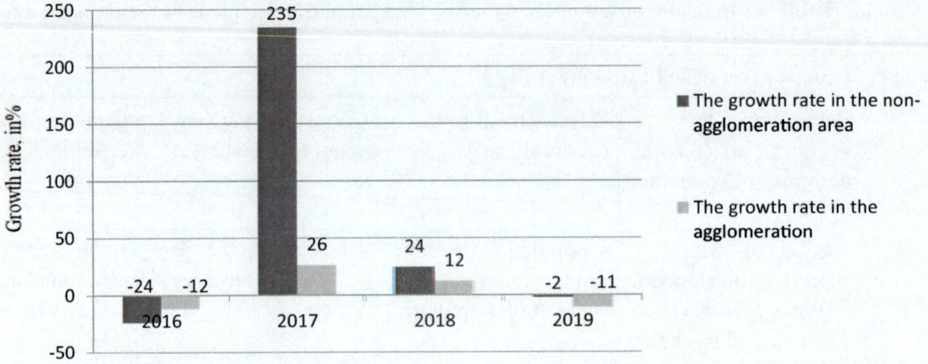

Fig. 1. Dynamics of the growth rate of investments in fixed assets per capita on average in the STA and in the non-agglomeration area in 2016–2019, %

In 2016, the subjects of the non-agglomeration territory shipped, on average, their own goods per capita more than the subjects of the STA by 34%, in 2020 - by 27.4%.

At the same time, from 2016 to 2018, the growth rate of this indicator for the non-agglomeration area is higher than for the STA (Fig. 2). In 2019–2020, the situation has changed, the intensity of growth in the volume of goods shipped by subjects of the non-agglomeration space has slowed down.

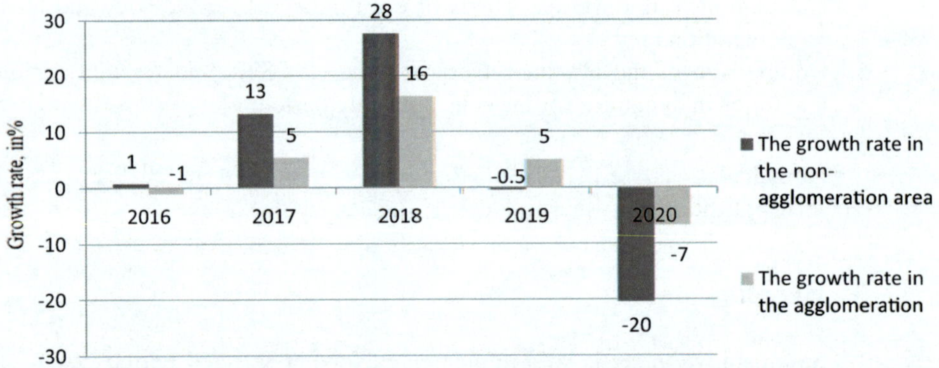

Fig. 2. Dynamics of the growth rate of shipped goods of own production per capita on average in the STA and in the non-agglomeration area in 2016–2020, %

The population density in the STA is extremely high. In 2016, it was 6.8 times higher than the population density of the territory outside the borders of the STA (533 and 78 people per 1 km^2, respectively). By 2019, this gap has not changed (531 and 77 people per 1 km^2, respectively).

The dynamics of the migration balance per 1000 inhabitants of the territory in 2016–2019 shows that the population decreases annually in the non-agglomeration territory, while the opposite situation develops in the STA (Fig. 3). In 2016, the migration gap between the two types of territories was 13 people per 1000 inhabitants, in 2019 – 7 people.

The analysis of the dynamics of the length of roads showed that in 2016 there were 1.86 km of roads per 1 km^2 of the STA, while in the non-agglomeration space it was 0.71 km (a gap of 2.6 times). In 2019, the gap slightly decreased to 2.4 times.

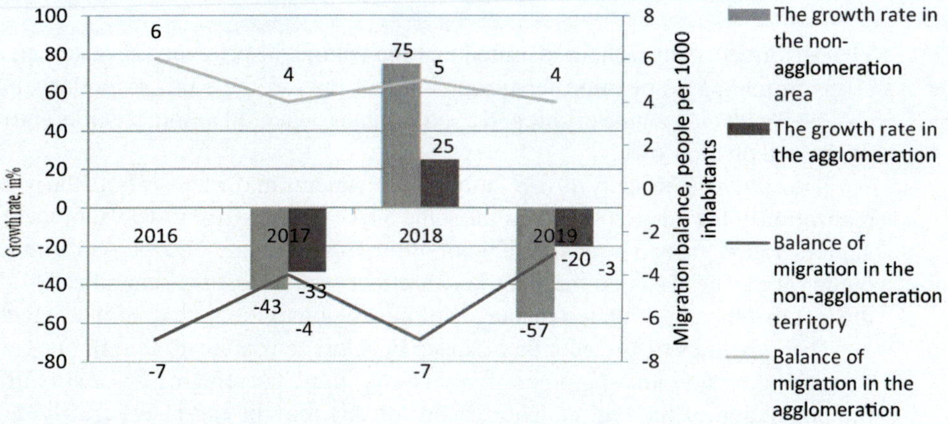

Fig. 3. Dynamics of the growth rate of the migration balance per 1000 inhabitants on average in the STA and in the non-agglomeration area in 2016–2019, %

The average monthly nominal salary in 2016 in the STA amounted to 27,247.7 rubles, in the non-agglomeration territory –23,845.0 rubles (a difference of about 15.2%) (Fig. 4).

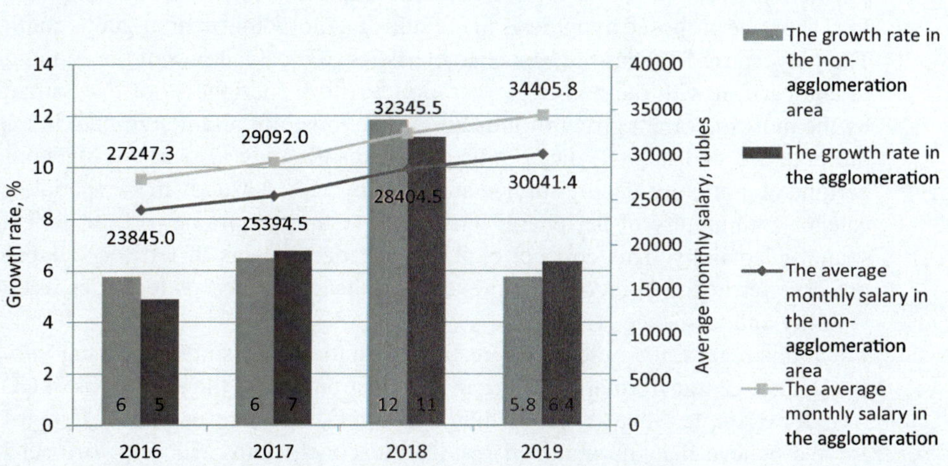

Fig. 4. Dynamics of the growth rate of the average monthly nominal accrued salary on average in the service station and in the non-agglomeration area in 2016–2019, %

In 2019, this ratio is: agglomeration – 34,405.8 rubles, non-agglomeration space – 30,041.4 rubles (a difference of 14.5%). In 2016–2018, the rate of salary growth in the non-agglomeration area is higher than in the agglomeration. Since 2019, the intensity of growth in two types of territories is slowing down, ratio has become the opposite.

4 Discussion and Conclusion

The performed analysis demonstrated that the volume of per capita investments in fixed assets in non-agglomeration territories is more than in the STA, a similar situation is observed with the volume of shipped goods of their own production, it works and services performed on their own.

The results of the analysis in relation to investments may indicate both the insufficient realization of the integration potential of the STA, and the advanced development of non-agglomeration growth points capable of attracting investment flows. As for the average output, it can be assumed that there is an effect similar to "deglomeration". It leads to the fact that the concentrated resources of the agglomeration are used inefficiently [12].

The dynamics of the economic space development indicate that the population is decreasing on the non-agglomeration territory, the STA remains the main attractor of the population of the region. The reason for this may be the insufficiently developed social infrastructure of the peripheral space of the Samara region, which is characterized by a limited set of services provided to the population and their rather low quality. The poorly developed transport infrastructure outside the agglomeration space increases the transaction costs of local producers and continues to slow down the development of the periphery. Low salaries in the non-agglomeration territory, compared with the average for the STA, do not create conditions for the formation of a high-quality human capital, and not only due to external factors that manifest themselves in dissatisfaction with the size of salaries, but also due to internal factors expressed in dissatisfaction with work and life. Thus, the proposed hypothesis of the study has not found unambiguous confirmation. The polycentric STA has the best characteristics of the dynamics of the economic space in comparison with the non-agglomeration territory (periphery) of the Samara region by the most important group of indicators. The most important argument in explaining the obtained and at first glance unexpected results may also be the phenomenon of peripheral immunity, deeply studied and fixed by M.Y. Kazakov in the spatial-economic category "immunity of peripheral territories" within the framework of his concept of spatial adaptalogy. This concept explains the mechanisms of territory adaptability to preserve territorial homeostasis, prevent the disappearance of localities, reduce their viability and resilience [13].

In this regard, it would be interesting to continue the study by taking into account new factors of transformation of urban spatial structures in the conditions of COVID-19 (see, for example, Stuart S. Rosenthal, William C. Strange, Joaquin A. Urrego) [14].

We believe that the obtained results and conclusions will help to form not only a new vision of the prospects for the development of the studied objects - the STA and the peripheral territory of the Samara region, but also a "new look" of both researchers and practical managers at the sustainable development of agglomerations and non-agglomeration spaces in other Russian regions.

References

1. Gainanov, D.A., Gataullin, R.F., Ataeva, A.G.: Methodological approach and tools for ensuring region's balanced spatial development. Econ. Soc. Changes: Facts Trends Forecast **14**(2), 75–91 (2021)
2. Minakir, P.A., Demyanenko, A.N.: Essays on Spatial Economics. Russian Academy of Sciences, Far Eastern Branch, Economic Research Institute, Khabarovsk (2014)
3. Giuliano, G., Kang, S., Yuan, Q.: Agglomeration economies and evolving urban form. Ann. Reg. Sci. **63**(3), 377–398 (2019). https://doi.org/10.1007/s00168-019-00957-4
4. Tao, M., Huang, Y., Tao, H.: Urban network externalities, agglomeration economies and urban economic growth. Cities **107**, 102–882 (2020)
5. Kolomak, E.A.: Spatial development of Russia in XXI century. Spat. Econ. **15**(4), 85–106 (2019)
6. Tsarev, A.I.: Notion in inner periphery and methods of its identification. Moscow University Bulletin, Series 5. Geography **4**, 33–42 (2019)
7. Kazakov, M.Yu.: Spatial-economic system definition of "Center/periphery" relation in the unity of integrative features. Econ. Yesterday Today Tomorrow, **9**(7A), 10–22 (2019)
8. Derudder, B., Meijers, E., Harrison, J., Hoyler, M., Liu, X.: Polycentric urban regions: conceptualization, identification and implications. Reg. Stud. **56**(1), 1–6 (2022)
9. Ouwehand, W.M., van Oort, F.V., Cortinovis, N.: Spatial structure and productivity in European regions. Reg. Stud. **56**(1), 48–62 (2022)
10. The Decree of the Russian Government: On the approval of the Strategy of spatial development until 2025. (No. 207-p dated February 13, 2019). Moscow, Russian Federation (2019)
11. Pavlov, Y., Koroleva, E.N.: Sources and results of knowledge generation about the Samara-Togliatti agglomeration. Bull. Samara State Univ. Econ. **5**(151), 19–28 (2017)
12. Pavlov, Y.: Development of urban agglomerations: Problems and solutions. Central Russ. J. Soc. Sci. **14**(5), 112–140 (2019)
13. Kazakov, M.Yu.: Spatial-economic systems "center-periphery": theoretical foundations, diagnostics of problems, strategic directions of development. Stavropol: AGRUS Stavropol State Agrarian University (2020)
14. Rosenthal, S.S., Strange, W.C., Urrego, J.A.: JUE insight: are city centers losing their appeal? Commercial real estate, urban spatial structure, and COVID-19. J. Urban Econ. **103381** (2021). https://doi.org/10.1016/j.jue.2021.103381. Accessed 31 Jan 2022

The Institutional Potential of Culture as a Tool to Ensure the Sustainability of Regional Society

Vadim Goncharov[1], Olga Kolosova[2], Larisa Tronina[3], Fatima Malukhova[4], and Valentina Ivashova[5](✉)

[1] North Caucasus Federal University, Stavropol, Russia
vgn1968@rambler.ru
[2] Krasnodar University of the Ministry of the Internal Affairs of the Russian Federation, Stavropol Branch, Stavropol, Russia
[3] Stavropol State Pedagogical Institution, Stavropol, Russia
[4] Kabardino-Balkarian State University named after H.M. Berbekov, Nalchik, Russia
fmalukhova@bk.ru
[5] Stavropol State Agrarian University, Stavropol, Russia
vivashov@mail.ru

Abstract. Balanced and progressive regional development is characterized by multi factors and the readiness of the society to resist internal and external challenges. Among the significant factors stabilizing the social situation is the institutional potential of culture as a branch of the social sphere. In studying the opinions of the region's inhabitants, the goal of the research was achieved. It is the definition of the institutional potential of culture, which acts as a tool to ensure the sustainability of the regional society. Furthermore, the sociological analysis of the population's involvement in cultural and leisure practices and satisfaction with the institutional conditions of functioning of sectoral organizations in the socio-economic system of the region was carried out. The high level of satisfaction of the population with cultural and leisure practices and readiness to spend their own free time in the institutional environment of culture is considered by the organizers of the study as full use of the potential and implementation of the established functions of cultural and leisure organizations in the region. For the regional society, the scale of involvement of the population in positive cultural practices provides its resistance to negative influences.

Keywords: Culture · Institutional potential of culture · Sustainability · Regional society

1 Introduction

The lack of attention and the residual principle of funding for a long time actualize the research interest in the current role of culture in providing a balanced socio-economic development of the region, meeting the cultural needs and demands of the population, increasing the attractiveness of the territory for living. The search for social stability in the regional context using providing cultural and leisure activities of different population

I. Samoylenko (Ed.): AMBT 2022, LNNS 582, pp. 140–147, 2023.
https://doi.org/10.1007/978-3-031-20803-4_15

groups is an actual field of research, which is confirmed by a brief review of publications in the designated subject area.

The institutional environment of a particular region has a significant impact on the industry's sustainable development. The author of the article, Negus K., shows the perspectives of the cultural economy from the point of view of influence on the personality and formation of favourable institutional conditions [1]. The institutional potential of culture is strengthened at the expense of economic mechanisms and provides a sustainable implementation of the regional cultural policy. That, in general, does not contradict our research strategy and confirms the relevance of the study of the designated subject area.

Diversity of cultural and leisure practices enriches the content side of family life and contributes to strengthening interpersonal relationships through collective cultural experiences. The authors of Ateca-Amestoy V., Ugidos A. conclude based on a large sample of married couples [2]. Family stability has a positive impact on essential characteristics of the territorial society through the involvement of more residents in positive cultural and leisure practices, the formation and reproduction of consumer culture, an increase in resource flows of the cultural sector. In general, these positions are also relevant for studying the institutional potential of culture as a tool to ensure the sustainability of the regional society.

The preservation of the cultural and historical heritage of the region is an important task facing the cultural sector. It provides continuity of the society's spiritual, moral, and cultural development, enriching its potential for resilience to the challenges of time. The authors of the article Saryglar C.V., Ertine I.A., Liundup T.M. use the example of organizing the work of the National Library of the Republic of Tuva named after A.S. Pushkin to show the introduction of digital platforms of cultural services and the development of the institutional environment for the successful functioning of the cultural sector [3]. Such social projects, which are aimed at supporting the creative development of people of different ages and social groups, contribute to the sustainability of the regional society [10].

According to the authors Kuatova Z.Z., Buletova L.A., Makhmutova K.I., Nuridinova G.A., Suleimenova A.A., the analysis of young generations' needs, expectations, and preferences in the cultural sphere allows forming a holistic view of the development of stereotypes of leisure behaviour [4]. For our study, the critical conclusion is that the emphasis in the publication is made on the continuity in the leisure practices of young people and supports the idea of stability of the value-normative system of society. The negative effect of the destruction of rigid traditional forms of social control and high uncertainty of new moral boundaries is reduced. The institutional impact of education and culture helps to solve the emerging negative deformations of the spiritual potential of society.

The development of cultural intelligence as an opportunity for social learning is considered by the author of the article Pauluzzo R. [5]. Modern technological solutions are necessary to understand the vectors of development of the institutional potential of culture. On the one hand, these technologies ensure the relevance of services for consumers. However, on the other hand, they maintain the institutional boundaries of the industry and its targeting. The development of a new direction is carried out for the entire satisfaction of the cultural and aesthetic demands and contributes to the development of the social and creative activity of the population. This direction becomes a positive contribution to the preservation of sustainability of the regional society [6–9].

Thus, the brief review of the literature indicates the relevance of the study of the institutional potential of culture as a tool to ensure the sustainability of regional society and the diversity of approaches to its implementation [11–14].

2 Materials and Methods

The empirical part of the study was organized in the Stavropol Region (Russia) in October–November 2021 using the Google Forms method. Quota territorial sampling was carried out in 9 territories of Stavropol Region: cities - Stavropol, Mikhailovsk, Kislovodsk; districts - Apanasenkovsky, Georgievsky, Novoalexandrovsky, Petrovsky, Sovetsky, Shpakovsky.

A total of 1,020 people permanently residing in the Stavropol Region and over 18 years of age participated in the survey. The sampling corresponds to the sex and age structure of the population of Stavropol Region. In studying the opinions of the region's residents, the goal of the research was to determine the institutional potential of culture, which acts as a tool to ensure the sustainability of the regional society. The goal was achieved by analyzing the involvement of the population in cultural and leisure practices and satisfaction with the institutional conditions of the functioning of sectoral organizations in the socio-economic system of the region. The high level of satisfaction of the population with cultural and leisure practices and readiness to spend their own free time in the institutional environment of culture is considered by the organizers of the study as full use of the potential and implementation of the established functions of cultural and leisure organizations in the region. Moreover, for the regional society, the scale of involvement of the population in the positive cultural practices provides its resistance to negative influences. SPSS Statistics package (version 21) was used for processing the database. Statistical distributions provided the segmentation of the survey results by types of settlements, gender and age of the respondents.

3 Results and Discussion

The survey results of Stavropol Region residents show a high level of satisfaction with services in the field of cultural and leisure activities: delighted 48.0%, largely satisfied 37.3%. Only 3.9% of respondents said they were dissatisfied with their visits to cultural institutions, and 10.8% found it difficult to give a definite answer.

Significant improvements in the work of cultural institutions recently were confidently reported by 21.4% of the survey participants, some improvements were reported by 45.9% of the survey participants, according to 30.6%, the quality of work remained the same, and only 2.1% indicated deterioration of work. Therefore, in general, we can talk about the positive changes in the activities of the cultural institutions of Stavropol Region, aimed at improving the quality of services by the current demands of the population.

Half (50%) of the survey participants believe that cultural institutions have significantly restructured their work during the pandemic, 7.8% noted that the work has remained at the same level, and 42.2% were unable to give a definite answer to this question.

The majority of survey participants expressed an opinion about the successful development of digital services of service provision in cultural institutions (29.4% said that digital services are actively developing and 44.7% said about some development); only 3.5% said that there is no development of digital services in cultural institutions; 22.4% were unable to give a definite answer.

In the course of the survey, the residents of Stavropol Region assessed their satisfaction with the conditions of the provision of cultural services.

On average, the conditions of the provision of cultural services in the cultural institutions of Stavropol Region received a score of 3.8 on a five-point scale.

To a large extent, respondents were satisfied with the competence and level of professionalism of professionals (3.9 points), politeness, tactfulness and friendliness of the institution staff (3.9 points), the culture of service in the institution (3.9 points), the presence of information signs and signs on the doors of the premises (3.9 points), staff appearance (3.9 points), cleanliness of the premises (3.9 points), checkroom service (3.9 points).

There remains some dissatisfaction with the range of services and informing the population about their provision. Also, areas for improvement in the work of cultural institutions should be the design of premises, convenience of waiting areas for visitors and the cost of services provided.

Table 1 presents the comparative data on the satisfaction of different population categories with the quality of the cultural services provided in the Stavropol Region.

Comparative data of satisfaction with the quality of cultural services in the Stavropol Region by different categories of the population shows that the highest overall satisfaction is observed in such groups as urban residents (3.85 points), men (3.95 points) and in the age group under 30 years (3.93 points).

Among the analyzed categories of the population, the most significant dissatisfaction with the quality of cultural services is observed in the age group over 50 years old, which indicates, on the one hand, about insufficient attention to the needs of this age group, on the other hand, about the high level of expectations of this group.

Areas for improvement of the work of cultural institutions and local authorities, responsible for culture and leisure of the population, can be associated with such demands of the population as an increase in the number of places to have fun (25.5%) and get together for communication (16.7%); as well as with the quality of organization of leisure activities and work on involving people in cultural leisure (9.8%).

In the survey, the residents of Stavropol Region collected suggestions about how they wanted to improve their cultural level.

The participants of the survey associate the improvement of their cultural level primarily with self-education (58.2%); with more frequent visits to concerts, plays (56.0%); reading (40.7%); more frequent visits to exhibitions museums (37.4%); learning foreign languages (34.1%); learning to play musical instruments, musical literacy (26.4%); learning the history and culture of their people (24.2%).

According to the survey participants, the most promising and valuable are such amateur initiatives in the field of cultural leisure as the development of sports and health

Table 1. Comparative data on the satisfaction of different categories of the population with the quality of cultural services in the Stavropol Region, in points on a 5-point scale

Response options	Type of settlement		Gender		Age			Total
	Urban	Rural	Male	Female	Up to 30	31–50	Over 50	
1. Information on the services provided by the institution (availability of a stand, reference information, consultant, booklets and other promotional materials)	3,6	3,4	3,8	3,6	3,7	3,8	3,2	3,6
2. Decorations of the premises	3,8	3,5	4,1	3,6	3,8	3,4	3,3	3,7
3. Cleanliness of the premises of the institution	3,9	3,7	4,0	3,8	4,0	3,6	3,4	3,9
4. Appearance of staff	3,9	3,9	4,0	3,9	4,1	3,6	3,3	3,9
5. Politeness, tactfulness and friendliness of the staff	3,9	3,9	4,0	3,8	4,0	3,7	3,4	3,9
6. The competence and level of professionalism of specialists	3,9	3,8	4,0	3,9	4,0	3,6	3,5	3,9
7. Availability of information signs and signs on doors of premises	3,9	3,9	4,1	3,8	4,0	3,3	3,5	3,9
8. Checkroom service	3,9	3,8	4,0	3,8	3,9	3,4	3,5	3,9
9. Convenience of waiting areas for visitors	3,8	3,5	3,7	3,7	3,9	3,2	3,4	3,7
10. Technical equipment of the cultural institution	3,9	3,4	4,0	3,7	3,9	3,6	3,4	3,8
11. The cost of services provided	3,8	3,7	3,7	3,8	3,9	3,3	3,5	3,7
12. The Range of services	3,7	3,4	3,9	3,6	3,8	3,4	3,4	3,6
13. The service culture in the institution	4,0	3,7	4,1	3,9	4,1	3,8	3,5	3,9
Average score by groups of survey participants	3,85	3,66	3,95	3,76	3,93	3,52	3,41	3,80

clubs (31.5%), art and creative associations, studios (25.8%), environmental movement (19.1%), associations and clubs of technical creativity (7.9%) and others.

The solution of many issues of development of cultural and leisure sphere is connected with financing of the branch. According to the survey participants, the state, enterprises, organizations, sponsors, and arts patrons should finance cultural expenditures. To the population as much as possible cultural services should remain free of charge (an option "free of charge" was noted by 26,1% of the poll participants, and an option "to a greater extent free of charge" was noted by 40,2% of the poll participants) or equally paid and free of charge (as noted by 27,2% of the poll participants).

Survey participants were unanimous in their choice of communication channels for receiving information about cultural and leisure services offerings: 83.5% would like to receive such information through social networks; 74.4% through Internet advertising; and 28.9% through advertising stands in populated areas. Such channels of information about events in the sphere of culture and leisure such as local TV, radio, newspapers are in demand by a small population (from 2% to 10%).

The importance of the research search is confirmed in the results of published scientific works. Thus, according to Kanellopoulou C., in the period of crises, more importance should be given to public space, providing requests in the field of culture, such as creative self-expression of artists [16]. Cultural practices of artists, which are carried out in the open public urban space, show the possibilities of creative self-realization for many other categories of the population.

The authors of Lebedeva L.G., Orlova L.V. also talk about the importance of institutional support of culture in the context of public space. [17]. The authors emphasize creating a conflict-free environment for intergenerational socio-cultural relations and exchanges of subcultures, taking into account the needs of different categories of residents.

The population in certain institutional conditions becomes a creator of new types of leisure. The author of the article by Chakars M. shows how cultural institutions involve the population in cultural and leisure practices and help the creative self-realization of an individual. For example, for young people, the risks of involvement in deviant behaviour are reduced, which improves the quality of life, health, economic activity, demographics and thus contributes to the stability of the regional society.

4 Conclusion

Different approaches in determining the institutional potential of culture as a factor of the regional society's sustainability have shown the high relevance of the topic and the importance [19–22], which were confirmed in the empirical part of the study.

The survey results of Stavropol Region's residents indicate a high level of satisfaction with services in the field of cultural and leisure activities. Significant improvements in the work of cultural institutions have recently been confidently reported by 21.4% of survey participants, some improvements were reported by 45.9% of survey participants, according to 30.6%, the quality of work has remained at the same level, and only 2.1% pointed out the deterioration of work. In general, we can talk about the positive changes in the activities of the cultural institutions of Stavropol Region, aimed at improving the quality of services by the current demands of the population.

Half (50%) of the survey participants believe that cultural institutions have significantly restructured their work during the pandemic, 7.8% noted that the work has remained at the same level, and 42.2% were unable to give a definite answer to this question.

The majority of survey participants expressed an opinion about the successful development of digital services of service provision in cultural institutions (29.4% said that digital services are actively developing and 44.7% said about some development); only 3.5% said that there is no development of digital services in cultural institutions; 22.4% were unable to give a definite answer.

On average, the conditions for the provision of cultural services in the cultural institutions of Stavropol Region received a score of 3.8 on a five-point scale. However, there remains some dissatisfaction with the range of services and informing the population about their provision. Also, areas for improvement in the work of cultural institutions should be the design of the premises, the convenience of waiting areas for visitors, and the cost of the services provided.

The residents of the Stavropol region wanted to improve their cultural level through self-education (58.2%); more frequent visits to concerts plays (56.0%); reading (40.7%); more frequent visits to exhibitions museums (37.4%); learning foreign languages (34.1%); learning to play musical instruments, musical literacy (26.4%); learning the history and culture of their people (24.2%).

The wishes of the inhabitants of Stavropol Region for improvement of the work of cultural institutions are reduction of prices for cultural events (38,2%); improvement of advertising of events (37,3%); creation of digital leisure resources (30,4%); allocation of spaces for communication, co-working centres (30,4%); improvement of cinema repertoire (30,4%); development of amateur clubs (27,5%); development of new competencies of specialists of cultural institutions (26,5%); development of family forms of leisure (23,5%).

It is possible to speak confidently about the high institutional potential of culture and its positive influence on the stability of the regional society.

References

1. Negus, K.: Perspectives on cultural economy: personal, institutional, historical. J. Cult. Econ. (2021). Article in print
2. Ateca-Amestoy, V., Ugidos, A.: Gender differences in cultural and sports activities attendance: an intra-couple analysis. Appl. Econ. **53**(55), 6426–6438 (2021)
3. Saryglar, C.V., Ertine, I.A., Liundup, T.M.: Preserving and promoting the cultural and historical heritage of Tuva by means of library work: the case of the A.S. Pushkin National Library of the Republic of Tuva. New Res. Tuva (2), 130–144 (2021)
4. Kuatova, Z.Z., Buletova, L.A., Makhmutova, K.I., Suleimenova, S.N., Nuridinova, G.A., Suleimenova, A.A.: Spiritual development as a factor of professional and personal qualities of modern students. J. Intellect. Disabil. - Diagn. Treat. **9**(1), 90–96 (2021)
5. Pauluzzo, R.: The imitation game: building cultural intelligence as a social learning capability to boost SMEs' international performance. J. Small Bus. Enterp. Dev. **28**(3), 317–336 (2021)
6. Bertacchini, E., Venturini, A., Zotti, R.: Drivers of cultural participation of immigrants: evidence from an Italian survey. J. Cult. Econ. **46**, 57–100 (2021). https://doi.org/10.1007/s10824-020-09405-0
7. Mansourian, Y.: Information activities in serious leisure as a catalyst for self-actualisation and social engagement. J. Doc. **77**(4), 887–905 (2021)
8. Kim, M., Jasper, A.D., Lee, J., Won, H.: Work, leisure, and life satisfaction for employees with physical disabilities in South Korea. Appl. Res. Qual. Life (2021). Article in print

9. Baigozhina, D.O., Ibrayeva, E.E., Duisengazy, S.M., Takhan, S.S., Zhanysbayeva, A.P.: Modern trends in mediatisation of culture in a digital society. Int. J. Criminol. Sociol. **9**, 3187–3197 (2020)
10. Biondo, A.E., Cellini, R., Cuccia, T.: Choices on museum attendance: an agent-based approach. Metroeconomica **71**(4), 882–897 (2020)
11. Sarmiento, I., et al.: Bridging Western and Indigenous knowledge through intercultural dialogue: lessons from participatory research in Mexico. BMJ Glob. Health **5**(9), e002488 (2020)
12. Idriss, S., Atie, R.: Race in Australia's youthful urban leisure scenes. Ethn. Racial Stud. **43**(10), 1854–1871 (2020)
13. Ortega Gaite, S., Quintano Nieto, J., Llanos Martín, S.: Youth leisure and alcohol consumption: socio-educational responses based on a program of alternative measures in Palencia city [Lazerjuvenil e consumo de álcool: Respostassocioeducacionais de um programa de medidasalternativasnacidade de palencia]. Pedagogia Soc. **36**, 155–166 (2020)
14. Chien, S.-Y., Lewis, M., Sycara, K., Kumru, A., Liu, J.-S.: Influence of culture, transparency, trust, and degree of automation-on-automation use. IEEE Trans. Hum. Mach. Syst. **50**(3), 205–214 (2019)
15. Sexsmith, K.: Decoding worker "reliability": modern agrarian values and immigrant labor on New York dairy farms. Rural Sociol. **84**(4), 706–735 (2019)
16. Kanellopoulou, C.: Space in common: socially engaged art in the Athens of crisis. J. Greek Media Cult. **5**(2), 211–230 (2019)
17. Lebedeva, L.G., Orlova, L.V.: The city as a social space of interaction between generations. IOP Conf. Ser. Mater. Sci. Eng. **603**(4), 042062 (2019)
18. Chakars, M.: Free time is not meant to be wasted: educational, political, and taboo leisure activities among the Soviet Buryats of Eastern Siberia. Natl. Pap. **47**(4), 660–673 (2019)
19. Zhu, L., Wang, K.: Urban agricultural development mode based on functional differentiation | [基于功能分异的都市农业发展模式研究]. Nongye Gongcheng Xuebao/Trans. Chin. Soc. Agric. Eng. **35**(10), 252–258 (2019)
20. Benítez, C.M.S., López, Á.L.: Spatial effects of cultural thematization for recreation and tourism in the pedestrian cultural corridors in Mexico City's Historic Center [Efectosespaciales de la tematización cultural para la recreación y el turismo en los corredoresculturalespeatonalesdel Centro Histórico de la Ciudad de México1]. InvestigacionesGeograficas (98) (2019)
21. Firbank, L.: What is sustainable agriculture? Biochemist **40**(4), 4–8 (2018)
22. Chauveau, H.: Culture and leisure in family farming: the practices of young people in rural South Brazil [Culture etloisirsdansl'agriculturefamiliale: pratiques de jeunesruraux sud-brésiliens]. Mondes Dev. **182**(2), 69–86 (2018)

The Development of Social and Industrial Infrastructure of the Agro-Industrial Complex as a Significant Indicator of the Sustainable Development of the Industry

Angelica Baicherova(✉), Natalia Telnova, Galina Tokareva, Darya Sidorova, and Alexander Tenishchev

Stavropol State Agrarian University, Stavropol, Russia
corsta@mail.ru

Abstract. The sustainable development of sectors of the national economy in modern realities is a vector for ongoing economic and social changes, which are unchanged in the context of globalization and consistent market transformations. Therefore, the issues of studying the mechanisms for ensuring the sustainable development of a particular industry and the entire socio-economic sphere are relevant and necessary. This article discusses the problems of modern understanding and evaluation of the effectiveness of the functioning of the infrastructure of the Russian agro-industrial complex as a significant indicator of the sustainable development of agricultural production. To do this, we used an extensive methodological base based on the use of abstract-logical, monographic, economic-statistical, computational-constructive methods, factor analysis and the method of canonical correlations. On the example of the municipal districts of the Stavropol Territory, a methodology for assessing the relationship between the state of development of the non-productive sphere of the agro-industrial complex (its infrastructure), production potential and the efficiency of production activities is formulated. Based on the results of the assessment, we proposed a grouping of the municipal districts of the study region according to the level of interconnection between the production potential of the agro-industrial complex and its non-productive sphere. The proposed methodology makes it possible to determine the factors that constrain the intensive functioning of the production and social infrastructure, on the basis of which it is possible to develop directions for improving and balancing the production and non-production spheres of the agricultural industry. The conducted study made it possible to draw attention to the importance of not only a quantitative assessment of social and industrial infrastructure, but the importance of assessing the relationship between the state of development of the non-productive sphere of the agro-industrial complex (its infrastructure), production potential and the efficiency of production activities in order to develop further tools to improve the efficiency of the industry.

Keywords: Agro-industrial complex · Sustainable development · Infrastructure · Social · Production · Factors

I. Samoylenko (Ed.): AMBT 2022, LNNS 582, pp. 148–158, 2023.
https://doi.org/10.1007/978-3-031-20803-4_16

1 Introduction

In the modern sense, the sustainable development of the agro-industrial complex is oriented, first of all, to the achievement of strategic goals of an economic and social nature. An essential condition here is the use of resource- and energy-saving technologies, the emphasis is on the greening of the production process and the rational use of biological factors in agriculture. In our opinion, ensuring the conditions for sustainable development of the agro-industrial complex should be based not only on the production and resource potential of the industry, but also on the level of development and quality of interaction between the industrial sector of the agro-industrial complex and its non-productive component (infrastructure).

In modern economic literature, there is no single approach to the interpretation of the term "infrastructure", while if we analyze all existing approaches, they can be conditionally divided into two groups:

– as a system of general purpose and use industries, functioning to provide conditions for the development of "private capitalist" enterprises [2, 8];
– as a complex of engineering and technical facilities that ensure the continuous dynamics of energy, transport and information cargo, passenger, energy and information networks and flows, without which the successful functioning of modern economic enterprises is impossible [1, 3].

Consequently, the term "infrastructure", due to its theoretical uncertainty and extended understanding, has generated many theoretical interpretations. Based on the generalization of the considered interpretations [4, 9], as well as in the context of this work, we will understand the infrastructure as a set of systemic connections of industries and areas of economic activity aimed at creating a continuous and efficient functioning of production processes.

The modern practice of agricultural production is inseparably based on the development of its social and industrial infrastructure. In a simplified practical sense, the infrastructure of the agro-industrial complex (agro-industrial complex) is a set of industries, sub-sectors that ensure the continuous, efficient implementation of the production process.

Intra-sectoral transformations of the agro-industrial complex are focused on the social market system of an organic type, while all key areas and industries interact on the basis of self-regulating ties. At the same time, intracomplex regulation is carried out on the basis of direct links between producers and end consumers [1, 2, 4, 8, 11].

It is important to keep in mind that agricultural production, as such, carries the properties of an infrastructural sector of the national economy, since its final result is only a stage of resource provision for the processing and food industries. That is why certain functions of the agro-industrial complex of a technological nature have separated and function in close connection with the service, industrial and commercial sectors, we ultimately take this functional relationship as infrastructure [2, 4, 6].

Based on the analysis of existing theoretical approaches to the study of the essence of the term "infrastructure", we single out the main developments and functioning of the modern infrastructure of the agro-industrial complex:

- production isolation;
- complementarity, that is, the admissibility of alternative and additional combinatorial options for the structure and configuration of infrastructure elements, taking into account production features;
- variety of forms [1, 7, 9].

It is possible to distinguish the infrastructure sectors of the agro-industrial complex from the main production by a number of key features, which, in our opinion, include the fact that it is of an integral nature, since it includes and accumulates heterogeneous activities in its system, and secondly, the functioning of the infrastructure takes place due to the horizontally directed functional division of labor. The third sign is the absence of a direct connection with the factors of cyclicity and seasonality of production characteristic of agriculture.

Considering the variety of modern interpretations of the term "infrastructure", we note the limited focus of the problem and the depth of the study of this concept. In the context of agricultural production, a significant proportion of research is devoted to the functioning of the production structure in the agro-industrial complex. At the same time, as A.N. Kokin, despite the key importance of the production infrastructure, "… it is necessary to note an extremely small number of studies of the infrastructure sectors of specific industries" [5]. We agree with this point of view, since it is obvious that the management mechanisms and principles of functioning of infrastructure elements are differentiated not only at the level of groups (for example, innovation, financial), but also within general groups (water supply, logistics). That is why we directed this study to expand the depth of the concept of infrastructure and ways to assess the effectiveness of its functioning in the framework of agricultural production [10].

2 Materials and Methods

Given the lack of a unified theoretical approach to the definition of the term «infrastructure», one of the key research methods for us was the monographic method, which allowed us to study the variety of interpretations of the term «infrastructure», the features of its theoretical research in the industry context. This allowed us to update the need to develop a methodology for assessing the effectiveness of the functioning of the agrarian infrastructure. For this, various research methods were also used, one of which was the abstract-logical method based on the determination of the object under study and the consideration of its problems in a «pure form».

Given the close relationship between the production and non-production spheres of the agro-industrial complex, in order to determine the properties and extent of the influence of non-productive industries on the actual producing sphere, we have moved away from the abstract-logical method.

To compare the level of development of the non-productive sphere of the agro-industrial complex with the level of its production potential and the degree of its effectiveness. To do this, we used the method of factor analysis. The main difficulty with which the solution of these problems is associated is the difficulty of quantifying the relationship between the state of development of the non-productive sphere of the agro-industrial complex, production potential and the efficiency of production and sales activities. For such an assessment, one has to operate with a large number of heterogeneous primary indicators, which are practically impossible to present in a visible and suitable form for unambiguous conclusions using traditional methods - statistical groupings, index estimates, and others. In this regard, we «compress» the initial information and instead of a set of indicators that are not comparable in terms of units of measurement, we introduce into the analysis a limited number of aggregated factorial values that replace them. To do this, we use the method of factor analysis. The essence of this method lies in the fact that on the basis of correlation and regression analysis for an internally related set of primary indicators |m|, a generalized factor sign is established, which is a vector value of this set and reflects in a quantitative form not individual particular characteristics, but itself this symptom in its entirety.

To study the influence of a set of interrelated quantitative indicators affecting the same set of resulting indicators, we used the method of canonical correlations. The essence of this method in general terms is reduced to a set of procedures and traditional methods of correlation and regression analysis, which allow to study the influence of a set of interrelated quantitative characteristics (independent variables) that simultaneously and jointly affect not one, but the system of resulting indicators (dependent variables). The research problem is reduced to the search for a mathematical interpretation of such linear sets of values of independent $X = (x1, x2 \ldots xi)$ and dependent ones $Y = f(y1, y2 \ldots yj)$ and convey basic information about the structure and dependencies of the objects under study.

Thus, based on the analysis of theoretical materials, we formulated the problem of the lack of a unified methodological and methodological approach to assessing the effectiveness of the functioning of the agro-industrial complex infrastructure.

In this regard, we proposed for use the methodology developed by us for a multifactorial assessment of the level of development of the non-productive sphere of the agro-industrial complex (its infrastructure) based on the method of factor analysis, grouping and highlighting the key factor features.

On the basis of the grouping carried out, a mathematical model for the interpretation of factor indicators was built, which made it possible to obtain basic information about the structure and dependence of the objects under study.

Further, using the method of economic and statistical interpretation, we carried out the corresponding mathematical calculations on the materials of the municipal districts of the Stavropol Territory using the MS EXCEL software package.

3 Results

We will begin to present the immediate results of this study with the systematization of the functions and elements of the modern infrastructure of the agro-industrial complex.

For clarity, let's turn to the data in Fig. 1, where we made an attempt to visually reflect the composition and structure of the modern infrastructure of the agro-industrial complex.

As can be seen from the above scheme, the production and social infrastructure of the agro-industrial complex perform mainly a supporting function within the production

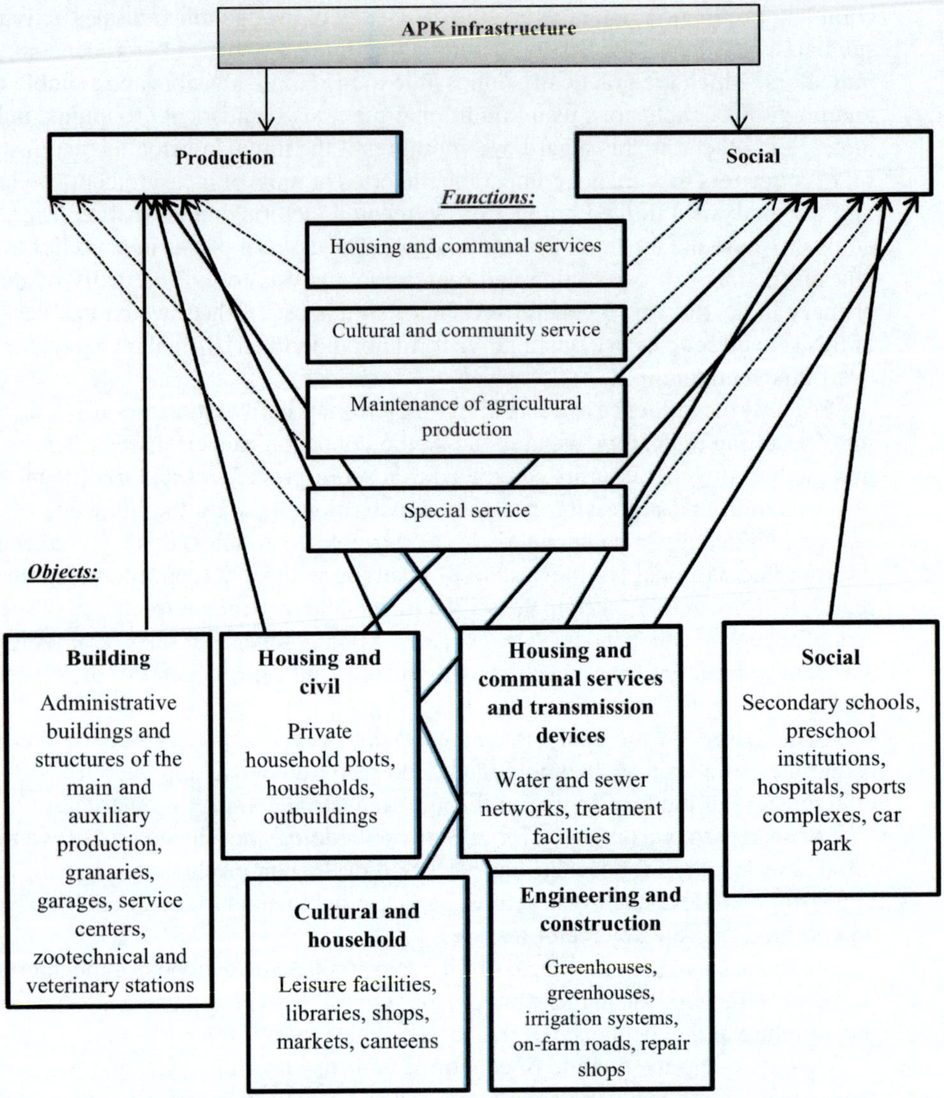

Fig. 1. Composition, structure and functional distribution of infrastructure facilities of the agro-industrial complex

and marketing processes of the industry. The depth and sustainability of infrastructure links, along with production and economic indicators, determine the sustainability and efficiency of the industry at all possible levels of research: from a municipal district to the level of a single state.

At the same time, we understand that the quantitative assessment of production and financial and economic indicators in agriculture is carried out on the basis of proven algorithms. Many methods and approaches have been proposed. Assessment of the level of development of the social and industrial infrastructure of the industry, at present, is reduced to a quantitative and dynamic calculation of the presence or absence of certain objects and supporting elements. So, for example, it will not be difficult to determine the number of general education schools, residential buildings, livestock buildings, granaries and other objects of the industrial and social infrastructure of the agro-industrial complex within a particular subject of study (village, district, region). The problem, in our opinion, is the lack of assessment methods aimed at correlating the available quantitative indicators with each other and with actual data on the productivity and efficiency of agricultural production.

Differences both in the level of production efficiency and in the degree of development of social infrastructure are reflected in the stability of the labor activity of agricultural workers and in the economic results of production.

The task is to develop directions for increasing the efficiency of the development of agriculture in the region by activating the development of the rural non-productive sphere.

Such a statement of the problem requires specific ways of solving it. In general, the essence of this approach in the development of this study is as follows: within the framework of the region under study, for each municipal district it is necessary to determine ("measure") the level of development of its non-productive sphere in comparison with the level of production potential and the degree of its effectiveness. The determination and further comparison of the districts of the region will make it possible to single out those of them that, according to the combination of the above characteristics, act as reference ones in relation to the rest. The indicators of these regions, extrapolated taking into account the prevailing trends in their changes within the established general standards, are taken as quantitative benchmarks for calculating the prospects for the development of the non-productive sphere in all other regions. At the same time, for certain groups of districts, a certain priority can be established for the development of individual elements of their non-productive sphere.

The calculations performed for the districts of the Stavropol Territory using this method made it possible to replace 54 initial indicators with seven generalizing factor values. The initial composition included indicators characterizing the state of agricultural production: its production potential and efficiency of functioning (Group 1), the demographic and labor potential of the village (Group 2), the level of development of the production sector of the region (Group 3). Let us consider in more detail each of the selected groups of indicators.

The first group, in our opinion, includes key production indicators, such as the area of agricultural land per farm and per worker, the yield of grain and fodder crops, the average annual milk yield per feed cow, the labor productivity of agricultural workers, the return on assets of agricultural production and the capital intensity of agricultural enterprises, the level of profitability of production, gross agricultural output per 100 hectares of agricultural land and per 100 hectares of arable land.

The second group consisted of the following indicators: the proportion of able-bodied people in the agricultural population, the progressiveness coefficient of the age structure of rural residents (the ratio of the number of children and adolescents to the number of persons of retirement age), the coefficients of mechanical and natural growth, the percentage of coverage of territories by Internet communications, the number of rural settlements per household, rural population density. These indicators reflect the level of development of social and industrial infrastructure within the studied object.

The third group of indicators covers the level of development of household infrastructure, this includes indicators of the provision of living space per rural resident, indicators characterizing the size of the satisfied and deferred demand of the rural population, the average length of roads, the coefficient of accessibility of medical and educational services.

As a result of the calculations, the system of the above primary indicators was summarized in the following factor indicators: the provision of the rural population with living space, trade and public catering enterprises (a1), the scale of development of the non-productive sphere of agricultural enterprises and its accessibility for residents (a2), the level of provision of the rural population with institutions health care, preschool education, the size of unsatisfied pent-up demand of the population (a3), the scale of concentration of production and the productive land fund of the economy (a4), the efficiency of agricultural production (a5), the progressiveness of the age structure and demographic movement of the rural population (a6), the scale of domestic current costs for research and development by type of work in the agricultural sector (a7).

Such a model can be calculated using the canonical correlation method.

With regard to the regions of the Stavropol Territory, such a calculation was carried out according to the following scheme.

The previously obtained factor indicators were divided into two groups – dependent and independent. The first group included values characterizing the state of the non-productive sphere (a1, a2, a3), and the second - the state and productivity of the production, demographic and innovation potential (a4, a5, a6, a7).

Several variants of various combinations of the first and second groups of factor indicators were considered, the assessment of which by statistical reliability would allow selecting the most significant model of the following interrelated linear combinations:

$$y = 0,278a_1 + 0961a_2$$
$$X = 6321a_5 - 0,419a_6 + 0,856a_7$$
$$K_{KaH} = 0,566$$

In the process of calculations, they dropped out as insignificant for this set of districts from the first linear combination (Y) factor a3, from the second (X) factor a4.

Substituting specific values for each region into linear combinations of the canonical correlation equation, we determine for any of them the values of the first (Y) and second (X) combinations and calculate the difference between them.

Next, we divide the two series of values of the calculated indicators for X and Y obtained for all regions into three intervals, in the first of which we attribute all the maximum, in the second average, in the third - their minimum values. A series of values of the difference between X and Y is also divided into three intervals. However, in this case, a slightly different distribution is established: all indicators close to zero belong to the middle group, which is defined as the most balanced in terms of Y and X. Indicators exceeding 0 - to the first (or third) group - with an imbalance in favor of Y; less than 0 - to the third (or first) group - with an imbalance in favor of X.

Thus, in accordance with specific indicators for each region X, Y and X-Y, it is possible to determine its place in the hierarchy of levels of development divided into three parts and the degree of balance between the production potential and the non-productive sphere. The districts that fell into the first group according to the value of X and Y and belong to the second group according to the difference between X and Y are among the reference ones, and the average indicators for this group of development of the non-productive sector, extrapolated to the future, are taken as quantitative guidelines for the development of non-productive spheres of the totality of the studied areas. The analysis of the regions of the Stavropol Territory carried out on this basis made it possible to distribute them into three enlarged groups. For a more visual acquaintance with their composition and structure, we turn to the data in Table 1.

Table 1. Grouping of districts of the Stavropol Territory according to the results of a quantitative assessment of the relationship between the production potential of the agro-industrial complex and its non-productive sphere.

Group	Districts	The degree of balance between the production potential and the non-productive sphere (X-У)	Source of imbalance	Directions for improvement
First group	Aleksandrovskiy	3,16	$a_2, a_4 \rightarrow min$	- an increase in the number of hospital beds by 21.1%; - increasing the productivity of the land fund; - increase in the number of technology parks up to 20
	Apanasenkovsky	3,18		
	Blagodarnensky	1,24	$a_7 \rightarrow min$	
	Izobilnensky	3,65	$a_4 \rightarrow min$	
	Ipatovsky	1,11	$a_1 \rightarrow min$	
	Kochubeevsky	4,12	$a_2 \rightarrow min$	
	Krasnogvardeisky	0,94	$a_6 \rightarrow min$	

(*continued*)

Table 1. (*continued*)

Group	Districts	The degree of balance between the production potential and the non-productive sphere (X-У)	Source of imbalance	Directions for improvement
	Mineralovodsky	2,12	$a_2, a_4, a_5 \rightarrow min$	
	Novoaleksandrovsky	4,14	$a_2, a_4 \rightarrow min$	
	Petrovsky	2,71	$a_7 \rightarrow min$	
	Shpakovsky		$a_4, a_7 \rightarrow min$	
Second group	Georgievsky	0,41	$a_2, a_4, a_5, a_7 \rightarrow min$	-increase in the provision of public catering places and institutions by 18.6%; - increase in the number of hospital beds by 19.1% - increase in the level of housing provision by 14.9%
	Kirovsky	0,13	$a_1, a_6 \rightarrow min$	
	Neftekumsky	0,78	$a_3, a_7 \rightarrow min$	
	Soviet	0,66	$a_2, a_4, a_5 \rightarrow min$	
	Arzgirsky	0,37	$a_2, a_4, a_5, a_7 \rightarrow min$	
	Levokumsky	0,85	$a_2, a_3, a_6 \rightarrow min$	
	Levokumsky			
Third group	Andropovsky	– 2,41	$a_2, a_4, a_5, a_7 \rightarrow min$	- increase in trade turnover per inhabitant by 28%, - extension of the length of roads by 12.1%, - stimulating the development of innovative infrastructure facilities; - activation of intra-industry cooperation processes
	Grachevsky	– 3,31	$a_1, a_3, a_6, a_7 \rightarrow min$	
	Turkmen	– 1,1	$a_1, a_2, a_4, a_7 \rightarrow min$	
	Stepnovsky	– 4,2	$a_2, a_4, a_5, a_7 \rightarrow min$	
	Kursk	– 3,6		

As can be seen from the above table, the level of development of the agro-industrial complex infrastructure among the regions of the Stavropol Territory is not uniform. The proposed array of data allows not only to interpret the results of the grouping, but also to identify the most significant factors that contribute to the formation of an imbalance in one direction or another.

4 Discussion

Thus, in relation to the Stavropol Territory, it can be concluded that the region has a high potential for the development of social and industrial infrastructure. At first glance, the obtained results of the grouping clearly correlate with the available statistical data reflecting the level of development of agricultural production. But the proposed methodology and the presented grouping allow us to form a broader view of the assessment of the existing problems of the development of the agro-industrial complex of the region. These problems are not always related to the yield or productivity of livestock. Sometimes problem areas are formed due to near-production factors, such as, for example, a lack of qualified personnel in a particular rural area. The study of the problems of the functioning of the industrial and social infrastructure using the proposed method allows us to evaluate its effectiveness not only from the standpoint of quantitative indicators (for example, 55 qualified veterinarians work in the district), but from the standpoint of the correspondence of these values to the real conditions for the functioning of agricultural production (for example, there are 22 veterinarians per 1000 head of cattle, while the district needs 34 doctors per 1000 head of cattle) This section includes a discussion of the results obtained during the study. You need to interpret and compare your findings with those of other authors or with your previous research findings.

5 Conclusion

The study allowed us to formulate a methodology for assessing the effectiveness of the functioning of the agro-industrial complex infrastructure in the context of its social and production components. At the same time, this assessment departs from the usual quantitative indicators, and is focused more on linking the production indicators of the industry's functioning with the objective conditions for the development of the non-productive sphere of the agro-industrial complex. The grouping of districts (enterprises) obtained as a result of using this method will allow using the data obtained to adjust and improve the conditions for the development of agricultural production as a key object of the agro-industrial complex and, as a result, will contribute to the sustainable growth of the industry. In this regard, we believe that the study has both scientific significance, since it is based on the development of a new approach to assessing the effectiveness of the functioning of the agro-industrial complex infrastructure, and practical significance, since the proposed research results can be used in assessing the investment potential of a particular economic entity and when developing targeted programs for the development of the industry.

References

1. Jochimsen, R.: Theorie der Infrastructur. Tuburgen (1966)

2. Fuller, S., Yu, T.H., Fellin, L., Lalor, A., Krajewski, R.: Effects of improving transportation infrastructure on competitiveness in world grain markets. J. Int. Food Agribus. Mark. **13**(4), 61–85 (2003)
3. Agibalov, A.V., Tkacheva, Y.V., Zaporozhtseva, L.A.: Improvement of the fi nancial management strategy for agricultural enterprises. J. Econ. Manage. Perspect. **11**(3), 1686–1696 (2018)
4. Gaiduk, V.I., Grishin, E.V., Bunchikov, O.N., Miroshnikov, D.M.: Factor environment for the development of industries of the industrial sector of the economy. Polythematic Network electronic scientific journal of the Kuban State Agrarian University. –No. 119. - rr. 1036–1052 (2016)
5. Kokin, A.N.: Improving the infrastructure support of the agro-industrial complex as the basis for its innovative development. Manag. Econ. Syst. **4** (2014)
6. Bunchikov, O., Shmatko, L., Yakovenko, Z., Riabchenko, A., Udovik, E.: State regulation of territorial development of agroindustrial region of southern Russia E-SdPTCONICIT – Espacios, vol. 41, no. 01, pp. 15–24 (2020)
7. Tarchokov, B.A., Fokichev, A.A., Khuranova, S.M.: Industrial infrastructure of the agro-industrial complex: essence and purpose. Humanitarian Soc. Sci. **6** (2007)
8. I.G. Ushachev, I.S. Sandu, V.G. Savenko. M. (eds.):Formation of the Innovation Infrastructure of the Agro-Industrial Complex: Problems and Prospects, p. 103 Capital Printing House LLC
9. Tastandiyeva, H.: Infrastructure development of agroindustrial complex. Вестник КазНУ. Серия Экономическая, [S.l.], vol. 107, p. 1 (2016)
10. Dos Reis, J.C., Bonacin, R., Correa, A., Jesus, K.R.E. Araujo, O.F.N.A.: Proceedings of 2018 IEEE 27th International Conference on Inclusion Technologies: Infrastructure for Shared Enterprises (2018)
11. Bornhofen, E., Ramires, T. Bergonci, T., Nakamura, L., Righto, A.: Agricultural Systems, 173 (2019)

Informational and Analytical Support for the Sustainable Development of Agribusiness Enterprises

L. I. Zinina(✉)

N.P. Ogarev Mordovian State National Research University, Saransk, Russia
zininali@mail.ru

Abstract. Digitalization and automation of the maximum number of processes is included as a recognized need in the development strategy of the largest agro-industrial domestic and global companies. The work is devoted to the study of problem areas and processes for developing an IT strategy for an agro-industrial enterprise in the context of the multidimensional interrelationships of the components of the business environment and the absence of explicit correlations between them. The paper reveals the essence of the IT strategy development model as the main tool for systematic impact on the enterprise architecture in its life cycle. As a result of the study, the goals and objectives of the business implemented with the help of information technologies are formulated, long-term directions for the development of information technologies and the functions of the IT department are determined, conceptual approaches to the implementation of strategic tasks of information management are reflected on the example of Teplichnoye JSC in the Republic of Mordovia. Based on the resource assessment of the current state of information technology and the IT department of an agribusiness enterprise, proposals were made to improve IT management, develop and implement an IT strategy.

Keywords: Digital transformation · Sustainable development · IT strategy · Enterprise architecture · Information management · IT strategy components

1 Introduction

The concept of sustainable development is built on the basis of the unity, interconnection, effectiveness of economic, industrial, social and environmental processes. It should be focused on the priorities of innovation management, providing for objective information technology solutions. The development of agribusiness has reached a level of maturity in which modern technologies are increasingly used to receive data from satellites, sensors, from operational and transactional systems. This increases both the amount of data and the need for high-quality processing and reliable conclusions that can be relied upon when making decisions. As a result, there is a demand for industrial analytical systems and advanced analytics. In the context of the digital transformation of the economy, the sustainable development paradigm in relation to agribusiness enterprises should

I. Samoylenko (Ed.): AMBT 2022, LNNS 582, pp. 159–170, 2023.
https://doi.org/10.1007/978-3-031-20803-4_17

involve the development of an information management mechanism and the development of an IT strategy, mainly within the framework of the enterprise's corporate strategy. This is all the more relevant in accordance with the requirements of the Sustainable Development Strategy until 2030 to achieve sectoral goals and objectives of an innovative nature. Therefore, specific information technology solutions are needed at all stages of its implementation. In this regard, we believe that there should be the development of methodological and practical approaches to the justification and implementation of a functional information strategy in the current conditions of digitalization.

Currently, the global information management community is trying to form the fundamental foundations in the development of an IT strategy. At the same time, it is advisable to name the developments of Danilin A.V., Slyusarenko A.I., Charaeva G.G., Eriashvili N.D., Abdikeeva N.M. and other domestic [1, 3–6] and foreign authors. At the moment, the theoretical foundations for the development and implementation of an enterprise informatization strategy are based on studying the practices of applying strategic IT management in historically effective organizations [7, 8, 10, 12, 14, 16]. This fact reflects the aspect of the narrowness of the approach and limits its application in the conditions of Russian realities in practice in most agro-industrial enterprises.. IT strategy cannot exist in isolation from the main business goals, on the other hand, it must also be aimed at IT aspects. To get the maximum effect from the costs of information technology (IT) in a company, it is necessary to develop and synchronize an IT strategy with the business goals of the organization [9, 11, 13, 15]. In this regard, the purpose of the article is to develop proposals for the formation of information and analytical support for the sustainable development of agribusiness enterprises in the context of the development of information and strategic management tools, in particular IT strategy.

2 Components of IT Strategy Development

We propose to consider the structural and functional content of the IT strategy development model on the example of an agribusiness enterprise (scenario approach) according to the following main components: block 1 (goal, constraints, approaches), block 2 (relationship with business strategy), block 3 (organization of IT spheres), block 4 (target enterprise architecture), block 5 (information integration), block 6 (IT infrastructure), block 7 (IT resource management system), block 8 (organizational changes), block 9 (target interaction), block 10 (sourcing), block 11 (financing), block 12 (transition plan to the target IP architecture), block 13 (options and risks), block 14 (selection of projects and software products).

Block 1: Goal-setting should be focused on the development of a specific version of the IT strategy, which provides for the use of modern information systems and resources and contributes to the formation of competitive advantages of the enterprise, which will allow solving a set of production and economic tasks at all levels of management, taking into account the features of the management system, strategic goals and prospects for the development of the enterprise.

Block 2: The problem of disunity between IT and business is the problem of the lack of development planning within the entire company. Attempts to improve the efficiency of the IT service, not correlating this activity with the overall business strategy, without involving other departments of the company, repeatedly proved the failure of this approach. The problem of disunity will disappear when the IT service ceases to develop separately from its company, when, considering the issue of IT optimization, top management will decide on the implementation of an ITSM project, subordinating the goals of the IT department to the goals of the company and determining the IT interaction scheme at the project planning stage with business.

Block 3: Assessment of the existing architecture of the enterprise (JSC Teplichnoye) was carried out according to the methodology developed by META Group, which proposed the use of a maturity scale consisting of five levels.

The mission of the enterprise is to provide the population with environmentally friendly fresh vegetables all year round. The company is constantly developing, mastering new technologies, establishing relationships with external partners, entering new markets. Emerging new strategic goals lead to an update of the enterprise architecture. The CEO who leads the company recognizes the importance of developing an enterprise architecture, but does not directly participate in it. The development of the enterprise architecture project is entrusted to the heads of departments, and then everything depends on their initiatives. As a result, not all of them become active participants in the process.

Active development is carried out only by the IT service according to the TOGAF methodology, according to which the enterprise architecture consists of business architecture, application architecture, data architecture and technological architecture. Some business units are only formally involved in the architecture development process.

The company has well developed internal standards and regulations. Implemented an integrated management system for compliance with GOST R ISO 9001–2015 (ISO 9001:2015) standards; GOST R ISO 22000–2007 (ISO 22000:2005).

Management, together with IT, defines the key requirements for the enterprise architecture. Interaction with vendors is carried out mainly by the IT service. The project planning cycle is continuous.

The technological architecture ensures the successful execution of business processes. Applications are classified according to their contribution to the business.

Planning and managing IT procurement is carried out in accordance with a specific architecture. Accounting for obsolete IT is organized. According to the conducted criteria analysis, the enterprise architecture is currently in a transitional state from a regulated level to a managed one.

Block 4: The structure and content of block 4 are presented in Table 1.

Table 1. Target architecture of the enterprise: main application systems in the implementation of business processes

Name of the structural element (block)	The main application systems used in the implementation of business processes			
	Buisness process	PP name	Users	Functional
Target Enterprise Architecture	1.Strategic management	Mozilla Firefox Browser	IT department, financial and economic service, supply and sales departments, agronomic service, protected ground workshops, energy service	Searching and viewing information on the Internet
	2. Production of products	Text editor LibreOffice Writer	IT department, financial and economic service, supply and sales departments, secretariat	Typing and editing texts
	3. Energy supply	Spreadsheet editor LibreOffice Calc	IT department, financial and economic service, supply and sales departments	Working with spreadsheets
	17. Strategic management	Mozilla Thunderbird	IT department, financial and economic service, supply and sales departments, agronomic service, energy service	Working with e-mail
	18. Production of products	ESET NOD32	All divisions of the company	Computer antivirus protection
	19. Energy supply	GIMP, Inkscape	Agronomic service, protected ground workshops, energy service	Graphic editor

(*continued*)

Table 1. (*continued*)

Name of the structural element (block)	The main application systems used in the implementation of business processes			
	Buisness process	PP name	Users	Functional
	20. Provision, storage and protection of information	1C: Accounting 8.3	Financial and economic service, supply and sales departments, warehouses, production workshops	Registration of business transactions, accounting and tax accounting, reporting
	21. Activity planning	FIREPLACE	Financial and economic service and personnel service	Calculation and calculation of wages, keeping records of personnel
	22. Purchase of goods and materials	sbis	Financial and economic service	Exchange of electronic documents with suppliers and buyers
	23. Sales of products	Circuit	Financial and economic service	Exchange of information and submission of reports to all state bodies (FTS, FSS, PFR, Rosstat)
		Client-Sberbank	Financial and economic service	Settlement and cash services
		ScalesSoft	Financial and economic service, warehouses, production workshops	Automatic reflection in the program "1C: Accounting" of finished products at the time of weighing
	24. Information security	RIC	OKS	Preparation of budget documentation

(*continued*)

Table 1. (*continued*)

Name of the structural element (block)	The main application systems used in the implementation of business processes			
	Buisness process	PP name	Users	Functional
	25. Production of vegetables in protected ground	Diana XP	Energy Service	Management and regulation of the operation of boilers in the boiler room
	26. Energy supply	Monitor	Agronomic service, protected ground workshops	Process control

Block 5: Information systems that are part of the portfolio of applied systems form and store the entire amount of information used in the company's activities. Settlement and cash services are carried out using remote banking technology (RBS). To do this, the enterprise has installed the Client-Sberbank program, in which the connection with the bank occurs via the Internet. The client program stores payment documents and account statements on the computer, allows you to generate and send payment orders, and receive information from the bank in the form of letters and account statements.

The technological program "Monitor" belongs to the number of specialized programs used at the enterprise. The network version of the Monitor program allows you to view and edit data via global and local networks. This function is very convenient when control over technological processes needs to be carried out from several places, for example, when an agronomist for irrigation and a technologist each control processes from their workplace.

Block 6: The network complex infrastructure is represented by the following components:

- server;
- structured cabling system;
- local computing network;
- uninterruptible power supplies (UPS) designed to protect the equipment and business processes of the company in continuous mode and at every particular second, as well as generator sets;
- mini automatic telephone exchange, combining external and internal telephone communication, modem, fax, collective answering machine and computer into a single system;
- network equipment (router, switches, etc.).

The specifics of the enterprise infrastructure is such that the direct use of basic network structures in its pure form is not possible. In this regard, the local area network of the enterprise is presented in the form of combinations of known topologies. Network switches are used to connect several workstations, respectively, to adapter cards.

Block 7: TO bring information systems and information technologies of corporate governance to a state that satisfies the needs of management bodies and personnel of the enterprise, as well as customers, it is required:

- Implementing a data culture (creating data as a service) so that employees can get the data they need without difficulty;
- transition to advanced analytics and artificial intelligence;
- development of technical support for the enterprise, including computers, server and network devices, office equipment and communications;
- training of enterprise specialists.

Block 8: The content of the block is shown in Table 2.

Block 9: Several options for interactions can be formulated:

- Business comes to IT with a task and expects proposals from IT to solve the problem;
- Business comes to IT with a ready-made solution for implementation;
- IT comes to Business with proposals and Business is ready to consider them.

Block 10: When choosing a contractor and an IT service provider, you need to pay attention to the following points. The executor must:

- deeply immersed in the client's business;
- understand the essence, the cause of the task/project;
- help the client formulate this task so that it is unambiguously described, achievable, useful;
- evaluate the solution of this problem, and not the cost of man-hours;
- solve problems and achieve improvements in the business of its customer.

Throughout the entire period of work, employees of the internal IT service should be provided with the necessary conditions for advanced training and self-development (draw up a training plan or create a training portal).

Block 11: The key point in choosing a vendor is the budget that the company is willing to spend on purchasing the product. The level of competence of the contractor depends on it. The source of financing is the company's own funds.

When drawing up a budget, you need to take into account not only the proposed prices, but also compare the estimates of specific tasks. Check with the vendor what exactly is included in this task.

Table 2. Management plan for the implementation of the implementation stage of the project to improve information systems and information technologies

Name of the structural element (block)	Tasks	Content	Responsible executor
Organizational improvement (organizational change)	Modeling a New Enterprise Database	Conducting an inventory and making corrections to the enterprise database Control measures for inventory	Technical Director; Department heads
	Improving the communication/communication system	LAN installation	Technical Director;
	Update of office equipment and computers	Identification of equipment replacement needs Selection of equipment supplier Procurement of equipment equipment Operational installation of new programs	Representative of the executing company
	Training of enterprise specialists	Control measures for updating and installing equipment	Technical Director;

With a fixed payment, start working with the contractor after the preparation of the terms of reference by the same contractor. During the creation of the TOR, the contractor's competencies and workflow can be assessed.

From the first days of the project, provide the contractor with access to the Gantt chart. Define checkpoints to check readiness and compliance with the work plan and schedule.

Block 12: The implementation of the following products will lead to continuous improvements in all business processes affected by them (introduction of correct and complete information, completion of tasks on time, reasoned shifts in time and personal responsibility for one's work). Necessary:

- firstly, in order to reduce the time for collecting information on labor costs and increase labor productivity, it was proposed to introduce the TimeControl Office time tracking system;
- secondly, in the process of strategic management, use the Podio online service to exchange relevant information and prompt decision-making;
- thirdly, in order to optimize the sales process, implement a CRM system that will eliminate the duplication of some functions and increase the efficiency of the department.

Block 13: When introducing new products, the following risks are possible when forming the amount of financing:

1. Unscrupulous or inexperienced contractors may offer to perform work at too low and therefore attractive price. Such conditions are most often agreed due to a limited budget. But they face the opposite effect: the cost of the project grows, the deadlines are stretched, the development of business processes is slowed down.

2. Unrealistically short project timeline. Most often, three parameters are important for business: speed, price and quality. But there is a rule: he agrees to a low price, he is unlikely to pay attention to the project, it is realistic to choose only two key parameters. Operational development from a qualified contractor will be expensive if he meets the stated deadline.

Block 14: AS part of improving the enterprise architecture for managing the company's activities, it is proposed to consider software products that allow you to create and control various projects, as well as application systems for employee time tracking and sales management (Table 3).

Table 3. Appropriate software products for current management tasks

Name of the structural element (block)	Name of the software product	Users	Functional
Selection of specialized software products (projects)	Time Control Office	Financial, economic and personnel services	Control of working hours of employees using contactless cards
	Podio	Administrative and management personnel	Setting tasks and tracking their implementation
	AmoCRM	Sales department	Sales Process Management

The introduction of new software products will lead to a change in the "information architecture" domain, namely the expansion of the existing database. Changes related to the improvement of the business architecture of the enterprise could not but affect the improvement of the technological architecture. With the introduction of new systems, it will be easier for IT professionals to serve business units and provide timely technology support for changes in the organization. For the uninterrupted operation of the proposed software products, it was recommended to introduce an additional physical server.

3 Recommended Functional IT Strategies and Implementation Steps

For the enterprise under study, the following functional IT strategies are appropriate, the essence and content of which are presented in Table 4.

Table 4. Recommended functional IT strategies in the conditions of the enterprise under study

Subject	Events
Consolidation strategy	Consolidation of information resources (databases), information systems and IT infrastructure
Strategy for creating a digital business platform	Planning the creation of logically unified information systems, data arrays, IT infrastructure
Business process informatization strategy	Determination of priorities for informatization of business processes, justification of the new functionality of information systems
Business Digital Transformation Strategy:	Significant increase in contribution to the business due to more complete automation of business processes and the introduction of new IT solutions
IT infrastructure development strategy	Planning the development of all technical means of the company
IT Services Development Strategy	Development plans for all IT services
IT Process Development Strategy	Plans for 1–3 years to improve all IT processes
IT budget planning	Planning for 1 year or more, IT costs, and IT and IS project plan
ICT project plan	Selection and planning of ICT projects for the next 1–3 years

The above scenario for developing an IT strategy can be adapted to various conditions of agribusiness organization. But at the same time, its implementation should be based on the use of automated information systems of a specialized nature, including application software (applications) for the formation of information and analytical support for each of the above structural and functional elements in the IT strategy model.

The development and implementation of an IT strategy is an objectively necessary process for the effective development of enterprises in the context of the digitalization of the socio-economic system. Its essence and content should be considered from the standpoint of:

a) creation and implementation of an information mechanism that can positively influence the competitiveness of an enterprise, information security requirements;
b) development of information technologies, IT resources, IT departments, IT infrastructure, etc.;
c) formation of functional and corporate strategies, integration of business and the IT industry [2].

We consider it expedient to implement the following stages of IT strategy development:

1. Justification of the mission, goals and objectives of the IT strategy.
2. Determination of the priority elements of the IT strategy and key performance indicators.
3. Analysis of external and internal conditions and opportunities in the development of the IT-sphere (STEP-analysis, SWOT-analysis, etc.).
4. Estimation of resources using BSC methodologies.
5. Formation of alternative scenarios for the development of the IT industry, taking into account the best practices.
6. Formation of an IT strategy in the form of a formalized document.
7. Determination of the IT budget, procedures and regulations, followed by the formation and distribution of IT functionality.
8. Monitoring the process of implementing the IT strategy, taking into account the requirements of the COBIT standard.

In view of the foregoing, in the process of forming an IT strategy, it is necessary to use the concept of a balanced scorecard, as well as the concepts of goal setting, key performance indicators, STEP analysis and SWOT analysis as tools for forming elements of an IT strategy and its alternatives. It should include analysis and evaluation of the IT architecture ("as is") and the target IT architecture ("as will be"), taking into account business requirements, key priorities, planning and implementation of business processes.

4 Conclusion

Summarizing the above materials of the study, it should be concluded that in this work, based on the conditions of the Concept of sustainable development, the need to develop the main tools of information management with an emphasis on the formation of an IT strategy for an agribusiness enterprise focused on the effectiveness of production and economic processes and on the priorities of innovative support is substantiated. Its content, priority components, specialized functional IT strategies, methodological aspects of implementing the IT strategy in the activities of an enterprise in the agrarian sector of the economy are determined.

The practical significance of the study lies in the ability to contribute to the implementation of proposals for improving IT management, IT strategy components, business processes of enterprise divisions, making objective information technology decisions.

The prospects of the study lie in the objective need to develop methodological and practical approaches to the formation and development of tools for strategic management in the information sphere based on ITSM with its subsequent adaptation in the conditions of large agribusiness companies in accordance with the objectives of their sustainable development.

References

1. Danilin A.V., Slyusarenko A.I.: Architecture and strategy. - M.: National Open University "INTUIT", p. 232 (2016)
2. Zinina L.I.: Formation of an enterprise IT strategy. In: Russian Economy In the Face of New Challenges: Materials of the All-Russian Scientific and Practical Conference (Saransk, 13–14 December), pp. 353–358 (2018)
3. Information management. In: Abdikeev, N.M., Bondarenko, V.I., et al.- Moscow: Infra-M, p. 400 (Textbooks for the MBA program) (2012)
4. Information management In: Charaev, G.G., Eriashvili, N.D., et al. - Moscow: UNITI-DANA, p. 359 (2012)
5. Lobanova, N.M., Altukhova, N.F.: The effectiveness of information technology. In: Yurayt, -M, p. 237 (2016)
6. Seletkov S.N.: Company information and knowledge management. In: Yurayt, – M., p. 208 (2017)
7. Brynjolfsson, E., Hitt, L.: Beyond computation: information technology, organizational transformation and business performance. J. Econ. Perspect. **4**(14), 23–48 (2000)
8. Dineva S.: Convincing your company leaders to invest in new technology. Harv. Bus. Rev. (2022)
9. Dow, K., Weidenmier, W.M., Shea, V.: Riding the waves of technology through the decades: the relation between industry-level information technology intensity and the cost of equity capital. Int. J. Account. Inf. Syst. **25**, 23–48 (2017)
10. Hagel, J., Brown, J.S.: Your next IT Strategy. Harv. Bus. Rev. **79**(9), 154–158 (2001)
11. Kearns G., Lederer A.: Alignment of information systems plans with business: the impact of competitive advantage. In: Proceedings of the Second AIS Conference, pp 840–842 (1997)
12. Laan, S.: Infrastructure Architecture - Infrastructure Building Blocks and Concepts, 2nd edn. Lulu Press, USA (2017)
13. Mutisari, R., Nugroho, C.P., Aprilia, A.: Information technology in improving agricultural marketing. Russ. J. Agric. Socio-Econ. Sci. **12**(96), 23–48 (2019)
14. Reich, B.H., Benbasat, I.: Factors that influence the social dimension of alignment between business and information technology objectives. MIS Q. **24**(1), 81–113 (2000)
15. Vacca, J.R.: Cyber Security and IT Infrastructure Protection. Syngress, Boston (2013)
16. Zhou, N., Zhang, S., Chen, J., Han, X.: The role of information technologies in firms resource orchestration process: a case analysis of China's «Huangshan 168». Int. J. Inf. Manage. **37**, 713–715 (2017)

Human Capital Development in Rural Areas from the Standpoint of the Sustainable Development Goals (SDGs)

O. N. Kusakina(✉) and S. V. Sokolov

Stavropol State Agrarian University, Stavropol, Russia
kusolga@list.ru

Abstract. The implementation of the concept of sustainable development affects different areas of society, in which the needs of the current situation are met, but the needs of future generations are not compromised. In this regard, it is particularly relevant to study the factors of achieving sustainable development goals, the key of which is the human capital of rural areas. This article describes the objectives of achieving sustainable development goals in their interrelation and interdependence with the elements of human capital of rural territories. We consider the directions and mechanisms of solving the tasks of forming human capital adequate to the processes of sustainable development, presented in the State Program of the Russian Federation "Integrated Development of Rural Areas". Furthermore, we reveal the problems of rural residents on the basis of a survey of rural residents in the southern regions of Russia. Finally, we substantiate trends in the development of modern agriculture and directions of human capital formation in the conditions of sustainable development of rural areas.

Keywords: Human capital in rural areas · Sustainable development · Global sustainable development goals · Agriculture · Sustainability drivers

1 Introduction

A major challenge today is the achievement of sustainable development goals in the economic systems of all countries, nationally and globally. The economic systems of different countries have their own characteristics, based on territorial-geographical, historical-cultural, and economic peculiarities, which determine the national identity. National identity, "is a specific unity of its elements, "meaningful attributes" shared by those who belong to a particular nation" [1]. But, despite this, there are global problems of present time, which require joint efforts to solve them. In recent years, humanity has defined the direction of the world development for the coming decades. The main concept of the future in the XXI century is sustainable development, the main statements of which are fixed in the conceptual documents of the UN. "The future we want" (2012) defines the prospects of humanity in the XXI century based on the concept of sustainable development, the basis of which should be a green economy [2].

I. Samoylenko (Ed.): AMBT 2022, LNNS 582, pp. 171–179, 2023.
https://doi.org/10.1007/978-3-031-20803-4_18

In September 2015, the 193 member states of the United Nations adopted a new sustainable development agenda, Transforming Our World: The 2030 Agenda for Sustainable Development [3]. The provisions articulated in this document are universal in nature, representing targets for all UN countries. The proposed Agenda for Sustainable Development is set out on the basis of common principles. The principle of universalism is manifested in the fact that all countries, irrespective of their national identity and economic development priorities, should pursue sustainable development goals. All countries should be involved in the sustainable development process, addressing the challenges together. In addition, the principle of inclusiveness is also manifested in the joint monitoring of processes taking place on the basis of real local data. The principle of connectivity and inclusiveness refers to the integrity of all 17 goals and the unity of the approach to their achievement. Inclusiveness determines the need for all sectors of society to participate in the implementation of the Sustainable Development Goals (SDGs). The principle of multi-stakeholder partnerships is also important to facilitate the implementation of the SDGs in all countries in the exchange of knowledge, experience, technology and financial resources [4].

There is no unambiguous understanding of the category 'sustainable development' in contemporary science, but in the context of achieving sustainable development goals it is worth agreeing that 'sustainable development (SD) refers to development that meets the needs of the present without compromising the ability of future generations to meet their own needs [5].

The achievement of sustainable development goals requires the fulfillment of a system of factors that are directly or indirectly related to the formation and development of different types of capital. Generalization of the existing points of view on the problem of the influence of different types of capital on sustainable development allows us to distinguish the following types: natural, productive, financial, social and human [5–7]. Natural, productive and human capital have a direct influence on the formation of material conditions for achieving sustainable development. However, given the multifaceted nature of the problem of sustainable development, all elements are interconnected, conditioning each other, determining the direction and results of development aimed at achieving sustainability of the system as a whole. Each type of capital has a direct or indirect influence on the other elements through the system of existing linkages, and at the same time the level of its development depends on the other types of capital. In addition, each has its own development mechanisms. All this makes the integrated system of sustainable development capitals multilevel and multifunctional. However, the specific integrator of this system for sustainable development is human capital, through which other types of capital are realised. The components of human capital such as individual abilities, knowledge, skills, health are related to the person himself and directly depend on the environment of its formation. In this regard, human capital in rural areas is a special type of capital with significant features. The existing scientific research in this area does not fully reveal the problems of human capital development in rural areas in the aspect of sustainable development.

The main objective of the article is to consider the relationship between the components of human capital of rural territories and the goals of achieving the SDGs, as the most important driver of the sustainability triad: economic, environmental, and social.

The article is based on the hierarchical subjective, structural-component, and integrative approaches to the consideration of human capital in rural areas as an object and a factor of sustainable development based on the use of statistical and verbal-communicative methods.

2 Informal Interpretation of the Problem

The fundamentals of modern human capital theory were laid by G. Becker, who studied social institutions (education, health care, etc.) that went beyond the scope of economic analysis, but significantly influenced the formation of human capital. H. Becker attached special importance to knowledge, production skills and motivation in the structure of human capital [8, 9].

It is now generally accepted that human capital is a factor of economic growth and qualitative development of socio-economic systems [10]. The multicomponent structure of human capital is determined by the subject approach, the subject-person, the bearer of human capital. Its characteristics, such as the occupation, the availability of necessary infrastructure for the formation of its elements, the generation of income, the impact of its activities on the environment significantly depend on whether it is a rural or an urban dweller. The environment of a rural person, a holder of human capital, has its own peculiarities. In Russia it is legally established that "a rural settlement is one or more rural settlements (settlements, villages, stanitsa, villages, farmsteads, kishlaks, auls and other rural settlements) united by a common territory, in which local self-government is carried out by the population directly and (or) through elected and other local self-government bodies" [11]. The development of rural ecosystems is largely determined by the symbiosis of different groups of factors: industrial and economic, natural, social, and ethical. Based on their combination and peculiarities of their manifestation in a certain territory, a special way of life is formed that distinguishes this rural area from others, including urban communities. This is reflected in the qualitative characteristics of the structural components of human capital in rural areas. We consider health, education, professional knowledge, cultural and personal potential of rural residents to be the key ones.

The conducted analysis of specific objectives within the framework of 17 Global Sustainable Development Goals has revealed direct and inverse links between them and different elements of human capital in rural areas. Carriers of human capital in rural areas play a key role in addressing the Sustainable Development Goals from the perspective of their access to unique resource and production potential, provision of food to the country's population and economic growth. Each of the Global Sustainable Development Goals is directly or indirectly linked to a particular element of human capital, which acts as a factor and outcome of this process. This consistency is reflected in the matrix of the sustainable development goals and the role of rural human capital in their achievement (Table 1).

The objectives of SDG 1 "End poverty in all its forms everywhere" apply directly to all holders of human capital in rural areas. It is only possible when significant resources are mobilized from a variety of sources, which is practically implemented by the relevant government programs. The current development of Russia's rural areas is based on strategic mechanisms set out in the State program "Integrated Development of Rural Areas" which is financed from the federal budget, the consolidated budget of the federal subjects of the Russian Federation and extra-budgetary sources. SDG 2 "Zero Hunger" implies increasing agricultural productivity and the efficient use of productive factors in rural areas; this requires improving the professional knowledge of agricultural workers, the development of the non-agricultural sector and the labour market in general. For example, a departmental project on "Promotion of Rural Employment" is being implemented through reimbursement of the costs of apprenticeship contracts concluded

Table 1. Matrix for integrating the sustainable development goals and the role of rural human capital in achieving them

SDGs	Objectives of the sustainable development goals	Elements of human capital in rural areas
SDG 1	Mobilization of resources from different development sources, programs and strategies to eradicate poverty in all its forms	Health, education, professional knowledge, individual capability
SDG 2	Increasing agricultural productivity and rural incomes, secure and equitable access to knowledge, land, other productive resources, and agricultural inputs; ensuring non-agricultural employment aimed at preserving cultural identity	Professional knowledge, individual capability
SDG 3	Ensuring universal health coverage	Health
SDG 4	Equal access to quality vocational and tertiary education, increasing the number of rural people with relevant skills for decent work	Education, professional knowledge
SDG 8	Progressive improvement in the efficient use of natural resources in rural areas and striving to ensure that economic growth is not accompanied by environmental degradation	Health, education, professional knowledge, individual capability
SDG 10	Progressive achievement and maintenance of rural incomes to urban levels as a precondition for access to health, education, and cultural services in rural areas	Health, education, professional knowledge, individual capability
SDG 11	Fostering positive economic, social, and environmental links between urban, peri-urban and rural areas	Health, education, professional knowledge, individual capability
SDG 17	Formulating more coherent policies for sustainable development	Health, education, professional knowledge, individual capability

with employees, costs of work placements and preferential loans for the construction of engineering infrastructure and residential facilities in rural areas.

Objectives of SDG 3 "Ensure healthy lives and promote well-being for all at all ages", aim to ensure universal health coverage and access to effective medicines for rural residents. SDG 4 "Quality Education" aims to provide rural residents with quality vocational and higher education and relevant occupational skills for employment. SDG 8 "Promote inclusive and sustainable economic growth, employment and decent work for all" aims to achieve economic growth through more efficient use of land and other resources in rural municipalities.

SDG 10 "Reduce inequality within and among countries" should achieve and sustain rural income growth that is significantly lower than urban growth. SDG 11 "Make cities inclusive, safe, resilient and sustainable" aims to maintain economic, social, and environmental linkages between urban, semi-urban and rural areas through improved infrastructure. SDG 17 "Revitalize the global partnership for sustainable development" is only achievable by maintaining positive economic, social and environmental linkages between urban and rural areas.

They include: bringing living conditions in rural settlements closer to urban conditions; increasing employment of the rural population; promoting the creation of new jobs by creating favorable infrastructure conditions for the development of agricultural and alternative activities; improving housing conditions for the rural population, as well as creating environmentally friendly living conditions in rural settlements, preservation, restoration and enhancement of human, cultural and natural potential of rural territories [12].

All above mentioned suggests that modern theoretical and methodological approaches to the study of the role of human capital in rural areas in solving the problems of sustainable development and the priorities of modern public policy do not contradict each other, and the research allowed us to consider in detail the objectives of achieving sustainable development goals in relation to the human capital of rural areas.

3 Formalisation of Factors for Achieving the SDGs

Russia's position on the agri-food map of the world is determined by the qualitative and quantitative characteristics of the human capital of rural areas, the resource and technological base of the industry, and the demand for agricultural products. Russia's total area of agricultural land is about 198 million hectares, or 4.5% of the world's agricultural land. By this indicator Russia ranks 5th in the world, behind China, America, Australia, and Brazil. One fifth of the world's freshwater resources are concentrated in Russia. Meanwhile, our country is the undisputed leader in agricultural land per capita, indicating significant potential to increase production and strengthen the country's contribution to global food security. Meanwhile, Russia ranks among the top 10 countries in terms of the value of agricultural production, which amounts to $85 billion, or 2% of global agricultural production.

In terms of gross output of certain products, Russia's agricultural sector is one of the largest in the world. Russia is the world's fifth largest producer of grain crops. The country is the second largest producer of sunflower oil and the third largest producer of cheese products.

Russia is among the world's top 20 agricultural exporters. There has been a three-fold increase in exports of food and agricultural raw materials in 10 years (Fig. 1). And according to the Russian Ministry of Agriculture, exports already exceeded a record $31 billion in 2021.

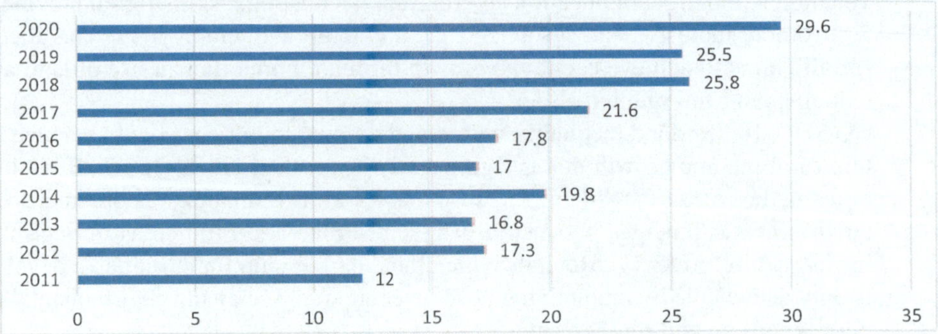

Fig. 1. Dynamics of food and agricultural raw material exports in Russia, USD billion

With a leading position in food exports, Russia has met almost all of the country's domestic demand for basic foodstuffs, making it possible to meet the thresholds of the Food Security Doctrine in key areas.

The agricultural sector employs 4.5 million people, which is just over 6% of the population employed in the Russian economy The number of the rural population and satisfaction with the rural way of life significantly depends not only on meeting the SDG targets directly related to addressing the food problem, but also on meeting environmental objectives and generally achieving economic growth.

According to the World Bank, 45% of the world's population live in rural areas, but in Russia this figure is significantly lower at only 25.2% of the total population. At the beginning of this year, Russia's rural population set a new anti-record, falling below 37 million (Fig. 2). At the same time, the rate of decline of the rural population is four times higher than that of the urban population.

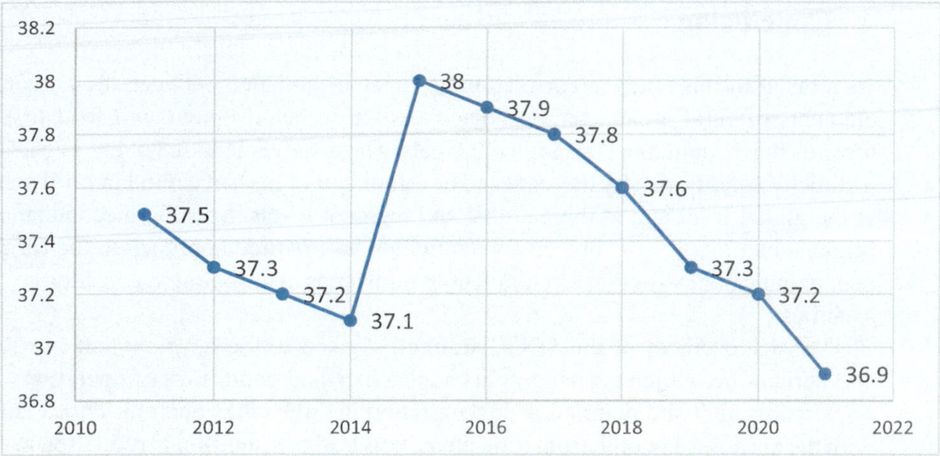

Fig. 2. Size of population residing in rural areas of the Russian Federation, thousand people.

To identify the problems of rural residents that cause their outflow from rural areas, the authors of the study conducted a survey of rural residents in the southern regions of Russia. The survey was conducted on the Internet using a standardised questionnaire created in the Google Forms service and the results are presented in Fig. 3.

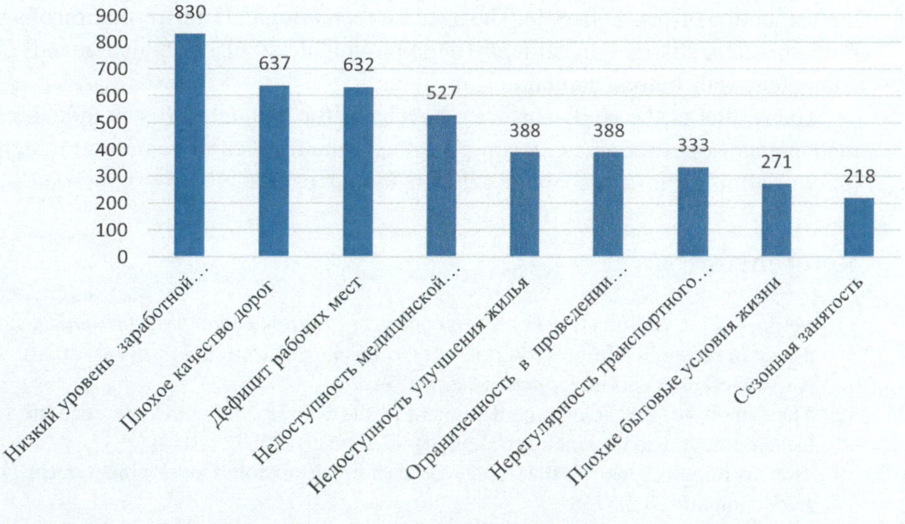

Fig. 3. Ranking of human capital problems of rural areas

The survey confirmed that rural residents consider low wages, poor road quality and limited labour market opportunities to be the key problems of life in rural areas. The level of workers' wages is a third lower than the average for all sectors of the economy.

4 Conclusion

As a result of this study, a correlation has been established between the human capital elements of rural areas and the objectives whose achievement will lead to progress towards the Sustainable Development Goals. The achievement of 8 of the 17 SDG targets is directly conditioned by the state and development of human capital in rural areas, both at the global level and at the national and regional levels. Specific mechanisms for the implementation of these objectives require a goal-oriented approach to the formulation of economic policy for the development of rural areas in the country, considering regional specifics.

The achievement of the SDGs is directly linked to the development of rural areas and agriculture, which is changing its appearance and conditions of operation.

Urbanisation and population concentration in large cities and megacities, combined with the outflow of people from rural areas, will lead to a significant reduction in employment in agricultural production and human capital. Global changes in the natural and climatic conditions of agriculture are contributing to a reduction in the planet's resource potential. According to a report by the International Panel on Climate Change, the average temperature on Earth could rise by 1.5 °C by mid-century, making many areas of traditional farming unsuitable for agriculture.

A clear trend is the transition to a new technological model, 'Agriculture 4.0', based on the global digitalisation and robotisation of processes, the spread of smart farming and integrated technology solutions, which requires a change in the approach to the formation of rural competencies. In the future, food production will increasingly depend on the intensification efforts of people. The need for technological transformation of agriculture fundamentally changes the structure of employment and places high demands on the key competences of human capital in rural areas.

The results of the study can serve as a basis for integrated sustainable development management of rural areas as complex socio-environmental-economic systems in the unity of all its factors, the key of which is human capital.

References

1. Velilaeva, L., Abdullaeva, N.: The concept of "national identity": theoretical and literary aspect of research. Univers.: Philol. Art Criticism: Electron. Sci. J. **6**(73) (2020). https://7universum.com/ru/philology/archive/item/9770
2. The future we want. Outcome document of the United Nations Conference on Sustainable Development. Rio de Janeiro, Brazil, 20–22 June 2012. UN (2012)
3. Transforming our world: the 2030 Agenda for Sustainable Development. https://docs.cntd.ru/document/420355765
4. The 2030 Agenda for Sustainable Development. https://www.unssc.org/sites/unssc.org/files/2030_agenda_for_sustainable_development_-_primer_russian.pdf
5. Sustainable Development indicators: Human Capital and accounting. The Academy of Economic Studies Bucharest
6. Goodwin, N.: Five kinds of capital: useful concepts for sustainable development. G.DAE Working Paper No. 0307 (2003)
7. Maryganova, E., Dmitrievskaya, N.: Human capital as a factor in sustainable development. Econ. Stat. Inform. **6**, 73–78 (2013)

8. Becker, G.: Human capital. The impact on earnings of investments in human capital, vol. 11, pp. 109–119. Economics, Politics, Ideology, USA (1993)
9. Becker, G.: Human behaviour. An economic approach, pp. 50–89. GU-HSE, Moscow (2003)
10. Shutaeva, E.: Multicomponent nature of the category "human capital" from the perspective of the concept of sustainable development. Int. Sci. J. Innov. Sci. **7**, 168–170 (2015)
11. Russian Federation. Laws on General Principles of Organization of Local Self-Government in the Russian Federation, Federal Law of 6 October 2003, No. 131-FZ (as amended by the Federal Law of 30 December 2021). http://www.consultant.ru/document/cons_doc_LAW_44571/
12. Russian Government Resolution No. 696 of 31.05.2019 "On Approval of the State Programme of the Russian Federation The Russian Federation's state programme "Integrated Development of Rural Territories Integrated Development of Rural Territories" and on amendments to certain acts of Government of the Russian Federation". http://pravo.gov.ru/proxy/ips/?docbody=&prevDoc=102142899&backlink=1&&nd=102558140. Accessed 24 Dec 2021

Author Index

I. Samoylenko (Ed.): AMBT 2022, LNNS 582, pp. 181–182, 2023.
https://doi.org/10.1007/978-3-031-20803-4